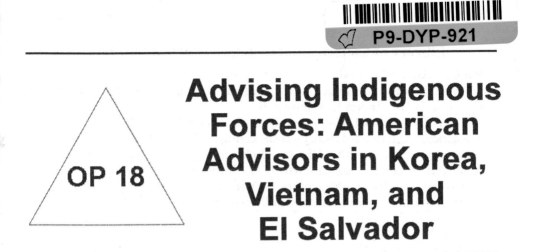

Advising Indigenous Forces: American Advisors in Korea, Vietnam, and El Salvador

OP 18

by
Robert D. Ramsey III

Combat Studies Institute Press
Fort Leavenworth, Kansas

Library of Congress Cataloging-in-Publication Data

Ramsey, Robert D., 1946-
  Advising indigenous forces : American advisors in Korea, Vietnam, and
El Salvador / by Robert D. Ramsey III.
      p. cm. -- (Global war on terrorism occasional paper 18) 1. Military
assistance, American--History--20th century. 2.   Military assistance,
American--Korea (South)--History--20th century. 3. Military assistance,
American--Vietnam (South)--History. 4. Military assistance, American--
El Salvador--History--20th century. 5. Military missions. 6. United States.
Army--Foreign service. 7. World politics--1945-1989. I. Title. II. Series.

  UA12.R35 2006
  355'.0320973--dc22

                                                        2006025616

CSI Press publications cover a variety of military history topics. The views expressed in this CSI Press publication are those of the author and not necessarily those of the Department of the Army or the Department of Defense.

A full list of CSI Press publications, many of them available for downloading, can be found at http://www.cgsc.army.mil/carl/resources/csi/csi.asp.

For sale by the Superintendent of Documents, US Government Printing Office
Internet: bookstore.gpo.gov Phone: toll free (866) 512-1800; DC area (202) 512-1800
Fax: (202) 512-2250 Mail: Stop IDCC, Washington, DC 20402-0001

ISBN 0-16-076722-9

Foreword

It has been said that the only thing new in the world is the history you don't know. This Global War on Terrorism (GWOT) Occasional Paper (OP) is a timely reminder for the US Army about the history we do not know, or at least the history we do not know well. The Army has recently embarked on massive advisory missions with foreign militaries in Iraq, Afghanistan, and elsewhere around the globe. We are simultaneously engaged in a huge effort to learn how to conduct those missions for which we do not consistently prepare.

Mr. Robert Ramsey's historical study examines three cases in which the US Army has performed this same mission in the last half of the 20th century. In Korea during the 1950s, in Vietnam in the 1960s and 1970s, and in El Salvador in the 1980s the Army was tasked to build and advise host nation armies during a time of war. The author makes several key arguments about the lessons the Army thought it learned at the time.

Among the key points Mr. Ramsey makes are the need for US advisors to have extensive language and cultural training, the lesser importance for them of technical and tactical skills training, and the need to adapt US organizational concepts, training techniques, and tactics to local conditions. Accordingly, he also notes the great importance of the host nation's leadership buying into and actively supporting the development of a performance-based selection, training, and promotion system. To its credit, the institutional Army learned these hard lessons, from successes and failures, during and after each of the cases examined in this study. However, they were often forgotten as the Army prepared for the next major conventional conflict.

These lessons are still important and relevant today. In fact, prior to its publication the conclusions of this study were delivered by the author to several of the Army's current advisory training task forces. We at CSI believe this GWOT OP can contribute significant insights to the Army as it works to prepare for and conduct its current advisory missions. *CSI—The Past is Prologue!*

Timothy R. Reese
Colonel, Armor
Director, Combat Studies Institute

Preface

This survey of American field advisors in Korea, Vietnam, and El Salvador began several months ago as a study on how the United States military had built foreign armies in Korea and Vietnam. The renewed interest in advisors caused a shift in the project. The final goal of this survey was to understand American field advisors—to determine who they were, to see how they prepared, to consider how they performed, and to discover what suggestions they had for improvements. The perspective of this survey is from the bottom, not the top. However, as always, what the top does and does not do affects how well those at the bottom are prepared, supported, and capable to perform their mission. The emphasis has been on understanding what was attempted and how the advisory effort evolved. I hope that this survey highlights issues that will be of use to current field advisors and to those who oversee advisory efforts.

No one does anything by himself. Among those requiring special thanks are Lieutenant Colonel Steven E. Clay for permitting me to do this project; Dr. Donald P. Wright for collecting the initial materials; John H. McCool for identifying the new materials in the Operational Leadership Experiences project; Combined Arms Research Library personnel Elizabeth Merrifield, Kathy Buker, and John J. Dubission for outstanding support in ferreting out materials in a short period of time; Martin J. Dockery for kindly providing a copy of advisor materials from Vietnam; Dr. William G. Robertson and Colonel Timothy R. Reese for supporting the project and for reviewing and improving the manuscript; and Betty Weigand for editing, instructing, and making it a better product. Without the support and effort of those above, as well as many others, this work could not have been completed in the time permitted. As always, responsibility for errors in fact or judgment rests with me alone.

For this survey, the spelling *advisor* vice *adviser* has been used in direct quotations for standardization.

Contents

Introduction

Do not try to do too much with your own hands. Better the Arabs do it tolerably than that you do it perfectly. It is their war, and you are to help them, not to win it for them. Actually, also, under the very odd conditions of Arabia, your practical work will not be as good as, perhaps, you think it is.[1]

—T.E. Lawrence, "Twenty-Seven Articles," 1917

T.E. Lawrence, Lawrence of Arabia, is occasionally read and frequently quoted today. As an earlier advisor to the Arabs, some consider his insights useful when facing current US military challenges—major advisory efforts, military and nonmilitary, in Afghanistan and Iraq. The article quoted above is frequently cited, but seldom with its last sentence. Yet this last sentence offers as much cautionary wisdom as the beginning. We must remember that Lawrence was not a soldier by trade. He had years of training and experience in the region. His linguistic concern was not with speaking the language, but with specifically speaking the local dialect. His unique situational understanding came from his academic background, his linguistic skills, his deep cultural understanding, his years in the region, his personal relationship with his counterpart, his understanding of what was and was not possible, his constant awareness of what was going on around him, and his unique genius. Only a few can or perhaps should be a Lawrence. However, he penned his "Twenty-Seven Articles" as suggestions for his co-workers, people more like us. By reading and thinking about the entire "Twenty-Seven Articles" (see appendix A), an advisor can grab nuggets worth consideration.

Not only is advising foreign indigenous forces a major focus of the United States (US) military today, but it will continue to be a major task in the future. During the on-going "long war" according to the latest Department of Defense *Quadrennial Defense Review Report*, US military forces must be able to "train, equip, and advise indigenous forces; deploy and engage with partner nations; conduct irregular warfare; and support security, stability, transition, and reconstruction operations."[2] To accomplish these tasks, advisors must have "language and cultural awareness to facilitate the expansion of partner capacity" and must "understand foreign cultures and societies and possess the ability to train, mentor, and advise foreign security forces and conduct counterinsurgency campaigns."[3] Fortunately, the US military has done this task before; unfortunately, it did not capture that experience in doctrine.

During World War II, United States military assistance missions provided equipment and training to the militaries of Great Britain, France, Iran, China, the Soviet Union, and Latin America. After the war, US Military Assistance Groups (MAG) or Military Advisory Assistance Groups (MAAG) provided equipment, training, advice, and assistance in Greece, the Philippines, China (Taiwan), Iran, Japan, and Korea. With the outbreak of war in Korea in 1950, US advisors faced the demanding task of providing advice during major combat operations—first, against North Korean and then Chinese communist forces. When the French withdrew from Indochina, the United States created the Military Advisory and Assistance Group, Vietnam (MAAG-V). By 1961, US advisors became involved in counterinsurgency operations with South Vietnamese combat units. By 1964, they advised province and district chiefs on pacification.[4] The advisory effort in South Vietnam continued until the US troop withdrawal in 1973. After Vietnam, US advisors continued to be active in many nations—Iran, Saudi Arabia, Latin America—working to improve their militaries and to maintain a friendly peacetime engagement. Longer-term efforts were provided in Iran and Saudi Arabia. In El Salvador, a small US Military Group (MILGROUP) oversaw a counterinsurgency effort for 12 years. Since then, Special Forces units have provided mobile training teams (MTT) for many small-scale, limited-duration training missions. As this brief overview shows, the US military has a vast and varied experience in advisory efforts.

To understand the basic challenges advisors faced and to determine what worked and what did not work, this survey looks at three US advisory experiences—Korea, Vietnam, and El Salvador. Korea, one of our first major advisory efforts, offers insights from a conventional combat experience working with a newly created, inexperienced South Korean army. Vietnam, our largest and longest effort, provides a combat and counterinsurgency experience working with a South Vietnamese army with previous combat experience. El Salvador permits a look at advising a limited, long-term counterinsurgency effort working with an army with no recent combat experience. Each of these conflicts is significantly different, yet the challenge confronted by the advisors—how to establish an effective working relationship with their counterparts to improve the host nation military effectiveness in addressing its security problems—was the same. In addition, the requirements to meet these advisory efforts came from the entire US military establishment, not just the US Army.

What did advisors do? How were they organized? How were they selected? How were they trained? What problems did they confront? How did their counterparts respond to their efforts? What suggestions did they offer? This survey investigates these questions. By focusing on field advisors—combat unit and host nation territorial teams immersed in the local situation with continuous face-to-face contact with the host nation officials—we can explore the lessons learned from these experiences and provide hard-earned suggestions to those in similar situations today. Reproduced in the appendixes are the suggestions these advisors developed for working effectively with their counterparts.

Notes

1. T.E. Lawrence, "Twenty-Seven Articles," *The Arab Bulletin*, 20 August 1917 [document online]: available at http://www.lib.byu.edu/~rdh/wwi/1917/27arts.html.

2. Department of Defense, *Quadrennial Defense Review Report* (Washington, DC: Department of Defense, 6 February 2006), 23.

3. Ibid., 23, 42.

4. Walter G. Hermes, "Survey of the Development of the Role of the U.S. Army Military Advisor" (Washington, DC: US Army Office of the Chief of Military History, Historical Report, 1965), 12–85. US Army official histories from the World War II period that address aspects of these advisory efforts include Richard M. Leighton and Robert W. Coakley, *Global Logistics and Strategy, 1940–1943* (Washington, DC: US Army Office of the Chief of Military History, 1955); Marcel Vigneras, *Rearming the French* (Washington, DC: US Army Office of the Chief of Military History, 1957); T.H. Vail Motter, *The Persian Corridor and Aid to Russia* (Washington, DC: US Army Office of the Chief of Military History, 1952); Charles F. Romanus and Riley Sunderland, *Stilwell's Mission to China* (Washington, DC: US Army Office of the Chief of Military History, 1953); Romanus and Sunderland, *Stilwell's Command Problems* (Washington, DC: US Army Office of the Chief of Military History, 1956); Romanus and Sunderland, *Time Runs Out in CBI* (Washington, DC: US Army Office of the Chief of Military History, 1959); and Stetson Conn and Byron Fairchild, *The Framework of Hemisphere Defense* (Washington, DC: US Army Office of the Chief of Military History, 1960).

Chapter 1

"A Much Tougher Job"[1]

KMAG Combat Unit Advisors in South Korea (1950–53)

I realize that I stand or fall with my counterpart. I share in
the credit for his successes and in blame for his failures.[2]

—"Ten Commandments" for KMAG Advisors, 1953

US Army forces occupied Korea south of the 38th Parallel in 1945
to disarm the Japanese Army and to assist in the establishment of a local
government. Soviet forces performed the same tasks in Korea north of
the 38th Parallel. By 15 August 1948, under President Syngman Rhee,
the government of the Republic of Korea (ROK) replaced the US Army
Military Government in Korea. The United States created a Provisional
Military Advisory Group (PMAG) to improve the effectiveness of the
Republic of Korea Army (ROKA). ROKA began as a constabulary to the
national police forces. When it became ROKA, its primary missions were
internal security and border defense. US advisors, working with ROKA
units, participated in counterguerrilla operations and monitored border
incidents. On 1 July 1949, the United States Military Advisory Group
to the Republic of Korea (KMAG), authorized 500 members, replaced
PMAG. KMAG was tasked to assist the ROK, within the limits of the
Korean economy, in developing its army, coast guard, and national police
by advising, assisting, and ensuring US military assistance was used effectively.[3] By the summer of 1950, ROKA had grown into a force of eight
light infantry divisions, each with varying levels of training, force structure, and experience.

KMAG During the Korean War

The North Korean attack on 25 June 1950 caught ROKA in the midst
of its building program. Five divisions had three understrength infantry
regiments; the remaining three divisions had two regiments. Only five
divisions had limited artillery units. ROKA had deployed four divisions
along the 38th Parallel, each with one battalion forward. Three divisions
were engaged in counterguerrilla operations in the south. Surprised, ill
equipped, and inadequately trained for the attack it faced, initial ROKA
resistance collapsed, particularly in the west. KMAG, assigned no
warfighting mission and caught between commanders, initially recalled
its unit advisors. Not all were able to comply. Some lieutenant colonel
and major division senior advisors and some captain and first lieutenant
regimental advisors remained with their ROKA units during the retreat.

But, the bulk of KMAG moved south to Taegu with ROKA headquarters. A KMAG detachment went to Pusan to facilitate dependent evacuation.

By 5 July, US Army forces from Japan began delaying the North Korean advance and reinforced what later became the Pusan perimeter. ROKA units on the east coast withdrew in good order, but most ROKA divisions required some reconstitution. Lieutenant General Walton H. Walker, Eighth US Army (EUSA) commander, took command of US forces in Korea on 13 July. With President Rhee's agreement, Walker assumed command of ROKA 4 days later. KMAG became a major subordinate command of EUSA with a secondary mission to maintain liaison between ROKA and EUSA. KMAG continued to advise ROKA on creating, training, and equipping units; EUSA controlled combat operations. KMAG advisor detachments with ROKA combat units responded operationally to EUSA, but administratively to KMAG. On 25 July, Brigadier General Francis W. Farrell took command of KMAG.[4]

Reconstituting ROKA combat units became the immediate task for KMAG. Only parts of five ROKA divisions existed after the retreat. All divisions lacked personnel, weapons, and equipment; however, ROKA divisions defended the northeastern portion of the Pusan perimeter. KMAG recommended and on 8 July ROKA created the ROK I Corps with two divisions positioned along the east coast, and on 14 July the ROK II Corps with three divisions positioned to the west of ROK I Corps. KMAG assigned advisors to assist the new ROKA corps commanders and their staffs in learning tactical and administrative skills the hard way, under combat conditions. Tasked in August to create the ROK 26th Regiment, Captain Frank W. Lukas, with two ROKA interpreters, contacted ROKA staff officers in Taegu. Working with local officials, within 2 days nearly 1,000 recruits had been drafted, formed into squads, then platoons, then companies, and then into two battalions. Those who "looked intelligent" were designated noncommissioned officers and platoon leaders. ROKA officers who helped recruit the soldiers became the company, battalion, and regimental commanders and staffs. Issued rifles, each soldier fired nine rounds of ammunition outside the city. Then the regiment, clad in a hodgepodge of civilian and military clothing, departed by train to the ROK 3d Division sector. This regiment, which entered combat less than a week after it was activated, received no formal training until April 1951.

On 9 August, Walker was authorized to increase ROKA strength immediately to any level he felt prudent. The EUSA plan called for doubling the number of ROKA divisions from five to ten with the first formed by 10 September and one a month thereafter to reach a total of ten. Within 6 weeks, EUSA reported three new ROKA divisions were organized and partially equipped and the remaining two were being activated. With additional divisions, ROKA activated the ROK III Corps in October. In November, after Seoul was recaptured, the KMAG chief asked an assistant G3 advisor, Major Thomas B. Ross, how long it would take to form a new ROKA division. Laughing at the several weeks' response, Farrell told Ross that he expected the ROK 2d Division formed by the next day. Ross went to the three personnel centers near Seoul, designated each a regiment, gathered recruits in groups of 200, designated each group a company, grouped companies into battalions, and grouped battalions into regiments. Company officers and noncommissioned officers were selected from those who looked intelligent. ROKA provided officers for battalion, regiment, and division commanders and staffs. Within days, the ROK 2d Division received rifles as it boarded trucks heading to a counterguerrilla mission. Units that were raised under such dire conditions and in such haste presented unimaginable challenges for ROKA commanders and their KMAG advisors.[5]

Fighting continued along the Pusan perimeter. Despite hasty reconstitution, lack of training, equipment shortages, and inexperience, ROKA divisions at times gave ground, but held their front. ROKA commanders, along with their advisor teams, performed well under demanding conditions. After X Corps attacked at Inchon on 15–16 September and recaptured Seoul by the end of that month, pressure on the Pusan perimeter diminished. EUSA, with the ROK I Corps and II Corps, exploited north to the 38th Parallel and then into North Korea. The ROK 1st Division served as part of US IX Corps. ROK I Corps advanced along the east coast so quickly that it secured the landing of US X Corps at Wonson on 26 October. In the west, EUSA advanced through P'yongyang, the North Korean capital, toward the Yalu River with the ROK II Corps on its right flank. On 26 October, a reconnaissance element of the ROK 7th Regiment, ROK 6th Division, reached the Yalu River, the first and only EUSA unit to do so. Unexpectedly, Chinese Communist Forces (CCF) attacked the ROK 2d Regiment late on 25 October and into the early morning of 26 October. The ROK 2d Regiment panicked, but 90 percent of its soldiers escaped. On 28 October, two additional ROKA regiments collapsed under CCF attacks, losing their vehicles and artillery. On 29 October, the CCF destroyed the ROK 7th Regiment—over 75 percent of the regiment and all of its advisors were captured or killed. The combined efforts of ROKA officers and their KMAG advisors could not overcome the basic weaknesses of the ROKA units. In less than a week, the CCF crippled the ROK 6th Division and weakened the ROK II Corps on EUSA's right flank. CCF intervention dramatically changed the EUSA situation.[6]

Anticipating another CCF attack, Walker repositioned forces with US I Corps to the west, US IX Corps in the center, and ROK II Corps to the east. The attack came on 25 November, and by noon on 26 November, the ROK II Corps front collapsed exposing EUSA to envelopment. Walker ordered a withdrawal. Attempting to disengage and move south, the US 2d Division suffered a severe blow. EUSA broke contact and moved to positions north of Seoul in South Korea. In the east, US X Corps and ROK I Corps moved to Wonson for an evacuation by sea that was completed by 24 December. On 23 December, Walker died from injuries suffered in a traffic accident and was replaced by Lieutenant General Matthew B. Ridgway. Ridgway, summoned from Washington, assumed command on 26 December, just as a major CCF offensive commenced. From west to east, the EUSA frontline was US I Corps with ROK 1st Division, US IX Corps with ROK 6th Division, ROK III Corps summoned from the south, ROK II Corps, and ROK I Corps on the east—eight ROKA and two US army divisions.[7]

During his 4 months in command of EUSA, Ridgway confronted CCF attacks, low US Army morale, and inherent weaknesses in ROKA. During the early 1951 CCF attacks, some ROKA divisions fought well, others not so well. KMAG advisors, the main source of information on ROKA units, proved unable to keep EUSA informed of the tactical situation. Breakdowns in communications between the front-line units and ROKA headquarters during combat were normal. In February, Ridgway directed the KMAG chief to personally accompany withdrawing ROKA units to ensure contact was maintained. In March, he warned all commanders against abandoning serviceable weapons and equipment. Lieutenant General James A. Van Fleet replaced Ridgway on 14 March when Ridgway moved to the Far East Command in Japan.

In April, the ROK 6th Division collapsed during a CCF attack. Despite the KMAG advisor's concern about the location of the division reserves, the ROK 6th Division commander placed them forward to reassure the morale of his forward units in anticipation of the attack. On 23 April, President Rhee asked Ridgway to support ten additional ROKA divisions. Ridgway bluntly informed President Rhee that ROKA needs were leadership and training, not personnel and equipment. Until ROKA demonstrated an ability to meet the demands of combat, Ridgway saw no point in raising more units. This meant Rhee, who strongly supported the actions of Ridgway and Van Fleet, had to remove incompetents and strengthen the ROKA officer corps. Until then, no expansion would occur. At the same time, a proposal from Washington for US officers to command ROKA units was rejected by Ridgway, Rhee, and the US Ambassador because it required a large number of US officers, it faced a language barrier, and it required authority to administer and discipline Korean forces. When the ROK III Corps failed to hold its ground in May, Van Fleet ordered its inactivation and the reassignment of its divisions to the ROK I Corps and the US X Corps. Over 24 KMAG advisors were killed or captured in the collapse of ROK III Corps.[8]

Ridgway assisted President Rhee in strengthening the ROKA. He directed Van Fleet to consider a special KMAG course for ROKA officers to address leadership deficiencies. Given the task, Colonel Arthur S. Champeny visited US Army schools and concluded that US Army courses were not suitable because of inadequate emphasis on the aspects of combat leadership required in Korea. On 22 July, in response to a Department of the Army (DA) inquiry into what was required to improve ROKA forces, Ridgway stated the first requirement was a professionally competent officer and noncommissioned officer corps that possessed a will to fight, an aggressive leadership ability, a professional pride, and a sense of duty. A second requirement was time. He believed it would take 3 years to improve the ROKA forces unless the war ended, then it would require 2 years. In July, Van Fleet appointed Brigadier General Thomas J. Cross as commander, Field Training Command. Within 3 months, each corps had a training camp for retraining ROKA divisions in corps reserve. A 9-week course progressed from individual weapons and tactics instruction to squad, platoon, company, and battalion. The tempo of combat operations, as usual, prevented most divisions from completing the entire course, but by late 1952 all of the original 10 ROKA divisions had completed 5 weeks of refresher training, and some had returned several times to accumulate up to 11 weeks total training. ROKA units benefited from this program by returning to the front with more skills, confidence, and esprit and by losing 50 percent less men and equipment than units without the training. The combined efforts of EUSA commanders, KMAG advisors, and President Rhee improved the combat effectiveness of the ROKA.[9]

The start of negotiations between the United Nations Command and China/North Korea in the summer of 1951 did not end combat operations. Nevertheless, a more stabilized front provided opportunities over the next 2 years for the improvement and expansion of ROKA. In October 1952, for example, the ROK 9th Division, suffering over 3,500 casualties, held its ground against 28 different attacks in a 10-day period by over 23,000 CCF soldiers. Late in 1952, a program to expand ROKA to 20 divisions and 6 corps would permit the replacement of US divisions in South Korea by ROKA divisions. However, things were never simple for KMAG. During the last half of 1952, it lost a large number of advisors to normal rotation.

Despite efforts by Van Fleet to channel additional personnel into KMAG, between losses of experienced personnel and increased requirements from ROKA expansion, KMAG remained, as normal, shorthanded. ROKA strength grew from 273,000 on 30 June 1951, to 376,000 on 30 June 1952, to 591,000 on 31 July 1953. By the July 1953 armistice, ROKA had become one of the largest combat experienced armies in the world.[10]

Combat Unit Advisors

KMAG grew from almost 500 personnel for an eight division ROKA in 1949 to a maximum strength in 1953 of 2,866 with 1,918 authorized; the rest were EUSA temporary duty (TDY) or attached personnel.[11] Because of the ROKA expansion during the war and other factors, KMAG personnel strength was always less than required. KMAG duties, in addition to advising combat units, included advising and assisting ROKA headquarters, its school system, its training base, and its logistics system. Consequently, most KMAG personnel were not in the advisor detachments at the ROKA division, regiment, or armor and artillery battalions.

Advisory Structure

Although the initial goal was to provide an American advisor to every division, regiment, and battalion commander, KMAG never had enough personnel for infantry battalions. Initially, a ROKA division had two or three infantry regiments with three battalions each and perhaps an artillery battalion. In 1951, divisions were authorized three regiments of three battalions, two artillery battalions, and one armor battalion. In June 1950, the average division advisory team, led by a lieutenant colonel or major, was five officers and three men. The regimental team, led by a major or captain, might be two officers. In March 1951, the new organization for the division advisory team was 21 officers and 11 men. Authorized at division were 2 lieutenant colonels—a senior division advisor and an assistant division advisor and executive officer; 5 majors—G1/G4, G2/G3, artillery, engineer, and signal; 2 captains—artillery and ordnance; and 11 enlisted men for mess, administration, communications, and maintenance. Each regiment was authorized one lieutenant colonel and one major for each battalion. Although authorized, personnel shortages prevented infantry battalion advisors throughout the war. By 1953, the division detachment was 44 officers and men, including a 16-man signal detachment from KMAG headquarters. Infantry regimental detachments were authorized two officers and two enlisted men; no advisors worked with infantry battalion commanders. Artillery and armor battalion advisory detachments were smaller. Many regimental advisors considered themselves understaffed to be an advisor, to do housekeeping, and to maintain 24-hour communications. Shortfalls occurred and many advisors did not meet the rank authorizations. KMAG regimental and battalion advisory teams—artillery and armor—were bare-bone, high-thrill outfits with no redundancy.[12]

Advisor Roles

Brigadier General William L. Roberts, the first PMAG and KMAG chief, established the "counterpart system." Each KMAG advisor was to work directly with his ROKA advisee—sharing office, activities, tasks, and problems. Unit advisors were to advise and assist—not command. Their impact was to come through personal influence, workable suggestions, and good guidance. Not only did advisors continue their peacetime advising and training functions, but with the war came the additional requirements to guide a counterpart—usually his superior by one to three ranks yet younger and less experienced—through the trials and tribulations of combat, to share life in the ROKA unit, and to maintain 24-hour liaison with EUSA units and KMAG headquarters. During the initial attack and reconstitution in the Pusan perimeter, some advisors clearly did more than permitted in responding to the grim situation. In any event, KMAG continued its "advise-and-assist, do-not-command" message. During the turbulent first year of the war, KMAG unit advisors faced the tricky task of ensuring that ROKA units stayed in the war and contributed to the effort. By 1953, KMAG issued its "Ten Commandments" (see appendix B) to its advisors. By then, it was clear that advisors still did not command, but they were responsible for the successes and failures of their ROKA units—they stood or fell with their counterparts.[13]

Advisor Selection and Tour Length

Before the war, KMAG service was not considered particularly desirable, important, or popular. In fact, KMAG duty was a routine assignment. Anyone could do it. Officers with the appropriate military occupational specialty (MOS), a need for an overseas tour, and the required rank found themselves in KMAG. No attempt was made to qualify personnel to provide appropriate advice to counterparts who outranked them by two or three grades. Often, advisors were junior company and field grade officers—willing and eager to do the job, but professionally weak. During the war, things did not improve. KMAG competed for scarce personnel resources with other US combat units in Korea. Except for brief periods in 1951 and 1953, priority for both quantity and quality of personnel went to US units. No matter how well an advisor did, he did not get the job opportunities or the one- or two-grade promotions many of his contemporaries did serving in US units. T.R. Frehrenbach summed it up, "Traditionally, a nation instructing another should send its best men abroad, traditionally, from Athens to the America of 1950, nations do not. There was little prestige, promotion, or hope of glory with serving with KMAG. The United States Army tended to forget these men. Most officers who could avoid KMAG duty did so, preferring to serve among their own troops, where food, companionship, and the chances of recognition were all considerably improved."[14]

Until it became important for ROKA units to replace the US divisions, little improved for KMAG. By 1953, some former battalion and regimental commanders, after 6 months in command in Korea, were assigned to KMAG as senior divisional and regimental advisors. Assignment criteria for other combat unit advisors did not change. Even an augmentation of KMAG late in the war by EUSA was a mixed blessing. The lack of qualified personnel was filled by large numbers of EUSA short-term personnel, many of whom were cast-offs in their units. While KMAG strength went up, the new personnel were of uneven quality and were available only for a short time. One adviser observed, "Getting worthless advisors relieved was easy. Poor advisors presented a greater problem." [15]

KMAG tour length prior to the war was 18 months unaccompanied and 24 months accompanied by dependents. During the war, tour rotation was based on a points system. US officers in combat units required 38 points to rotate. They received four points each month in line forward of regiment, three points each month in reserve between regiment and division command posts, and two points each month behind the division. KMAG regimental and divisional advisors required 40 points to rotate and received only 3 points each month. To KMAG advisors, this was another indication that their work was not understood or appreciated.[16]

Advisor Preparation and Training

KMAG advisors, much less combat unit advisors, received no special preparation or training. Before the war, a new advisor received a short orientation, met the KMAG chief, visited with the staff, attended a weekly staff meeting, and received an advisor handbook and a procedure guide for his reading. Then he went to his assignment. During the first 3 years of the war, a new advisor received, at most, a brief KMAG orientation and/or a short briefing by his immediate superior before going directly to his unit at the front. Chronic shortages and frequent turnover of personnel, combined with the chaotic tempo of back-and-forth combat operations and the expanding KMAG operational duties, meant that many unit advisors received little information on their mission or duties, much less the conditions under which they worked. As a new advisor commented, "The officer I replaced met me at the railhead (4 hours behind the division), turned his jeep over to me, and gave me directions to the division CP." Division and regimental advisors were often thrown into a make-it-happen situation with little guidance and limited support. They learned from their co-workers in the combat units and from experience. However, by the summer of 1953, things were more organized. KMAG conducted regular orientations, provided advisors with an *Advisor's Procedure Guide* that emphasized the twin duties of advising their counterparts and of providing accurate and timely reports to US commanders, and ensured each advisor received the "Ten Commandments" for KMAG advisors. Not surprisingly, during the war many combat unit advisors felt ill prepared for what they faced.[17]

Challenges of the Advisory Environment

Advisory duty in Korea was not typical military duty. Most officers considered being in combat with trained, equipped, and well-led US units a difficult and demanding task. The KMAG unit advisor faced that same challenge with an ill-equipped, inadequately trained, and inexperienced unit. In addition, he needed the professional skills and knowledge to advise a ROKA commander one to three levels above his rank or experience— captains and majors advised regiments, majors and lieutenant colonels advised divisions. An advisor's situational understanding was limited at best. He did not understand the culture, the language, the capabilities and limitations of ROKA units, his counterpart, nor US Army expectations. The 1953 *Advisor's Procedure Guide* offered a lot on what to do, but little on how to do it:

> In overcoming such obstacles as the language bar-
> rier, archaic beliefs, superstitions and a general lack of
> mechanical skills, the task of the Advisor has been an
> arduous one. The function of the Senior Advisor to a ROK
> regiment perhaps best illustrates the problems an Advisor
> faces. Living, working, fighting and training with a regi-
> ment, an Advisor must be acquainted with every phase
> of the regiment's operations. He must be abreast of the
> tactical and logistical situation. He must know the strong
> and weak points of the command and his subordinates. It
> is upon him that the regimental commander depends for
> knowledge that will teach him teamwork in the employ-
> ment of infantry, artillery, air, signal communications and
> armor in a combat operation and of the various services
> in support of the same. He must criticize their mistakes

without causing them embarrassment or "loss of face." He must teach them economy without seeming to deprive them of their needs. He must hold them to proven military methods and standards while still applauding their improvisation and, last but not least, he must do these things with a view toward building their confidence.[18]

How a division advisor, a member of a relatively small detachment, or how a regimental advisor, at most one of two officers, tackled the arduous task of overcoming cultural and linguistic obstacles and accomplished the advisory tasks was left largely to trial and error.

Understanding: Culture and Language

Americans and Koreans were different in almost every way—values, beliefs, social practices, religion, history, and language to name a few. Knowing what those differences were would have benefited KMAG advisors. Understanding the why of those differences would have been even better. The US Army and KMAG tended to expect the Koreans to understand and adapt to the Americans rather than to focus on what an advisor needed to know to work effectively with his counterpart. Some advisors even wondered if American customs and practices could actually replace those of Korean culture. The KMAG view that Korean cultural differences were "archaic beliefs, superstitions," not a different set of legitimate customs and beliefs to be understood in developing feasible recommendations, further reinforced the belief that the American way was the best way. All of this made it more difficult for most advisors to understand their Korean counterparts.[19]

What passed for cultural awareness was generally superficial, descriptive, and stereotypical. Advisors knew the concept of face was important, but few understood what the concept actually meant and why it was important. To most advisors, it was a reminder not to embarrass his counterpart. That many unknown cultural practices could embarrass a counterpart escaped their notice. The importance Koreans placed on respect, proper behavior, authority, and on a unit commander differed from US Army practice. Common Korean cultural faults acknowledged by most KMAG advisors included lack of initiative, failure to plan, inflexibility, no understanding of "the big picture," "lack of judgment," and "don't understand cooperation." That these judgments reflected an American cultural view remained hidden from most US advisors. The many US military concepts introduced by Americans, such as leadership expectations and the role that the commander performed, were just as alien to the Korean culture and practice as Korean concepts were to the Americans. In a 1953 study, none of the

14

KMAG officers surveyed considered the possibility that a good American leader might not be a good Korean leader because of different culture, practices, or standards. KMAG advisors knew they were "members of the best fighting organization in the world and . . . backed by the most highly developed industrial country in the world." Cultural understanding required not only empathy with Koreans, but also self-knowledge—understanding American cultural strengths and weaknesses.[20]

Language problems in Korea were commonly referred to as the language barrier. From the beginning, KMAG had no Americans who knew or understood the Korean language. By 1953, only one KMAG advisor could speak, read, and write Korean with any degree of fluency and only one other was fluent in Japanese. Prewar attempts to get KMAG advisors to learn Korean failed from lack of interest. An early KMAG chief actually considered making English the common language of the ROK security forces, even though most Korean soldiers were illiterate. Somehow, thousands of illiterate Koreans could learn English better and faster than educated Americans could learn Korean. To make an initial breach in the language barrier, ROKA commissioned university students with English language skills as interpreters and established an English language school for officers sent to training courses in the United States. Another problem, beyond getting people who generally understood the same word in two languages, was that words did not exist in Korean for many military terms and pieces of equipment. When there was no appropriate word, the item or concept was described as well as possible by the interpreter. For example, a machine gun, for which no word existed in Korean, became a gun that shoots very fast or a gun of many loud noises or any other way the interpreter chose to describe it. Such interpretations were inconsistent, inaccurate, and "ranged from the ingenious to the inadequate." A major effort to translate the *Dictionary of United States Army Terms* into Korean terminated with the outbreak of war. The shortage of Korean language translators and a lack of vocabulary created a language barrier that made communications difficult and created confusion and misunderstandings.[21]

Communication is critical to situational understanding; situational understanding is critical to useful advice. Lacking Korean language skills, KMAG unit advisors relied heavily on ROKA interpreters. Sometimes an advisor would pick up some Korean phrases or his counterpart might know some English. Koreans welcomed even a modest effort by KMAG unit advisors to understand them and their culture. Generally, none of the three—advisor, interpreter, or counterpart—knew all that he needed to communicate effectively without misunderstanding or confusion. ROKA interpreters were assigned to the unit. Their primary loyalty was to the ROKA commander, not to the advisor. Interpreters, commissioned from the university, were commonly referred to as schoolboys by their commanders. Without any military training, interpreters lacked a necessary understanding of military terms and concepts as well as colloquial English. The ability of the interpreter to understand, and then clearly translate not just the words but the meaning was rare. In addition, working effectively through a translator in a combat situation remained difficult and time-consuming.

Many advisors lacked even the basic ability to assess the reliability of the interpreter.[22] A study of KMAG advisors concluded: "From the circumstances there can be no question but that the utter dependence of KMAG advisors on Koreans for vital military information, and as a corollary the inability of KMAG advisors to obtain information directly through the use of the Korean language, hampered them in the accomplishment of their mission and resulted on numerous occasions in the unnecessary loss of territory and lives or wastage of ammunition." An advisor added: "For an individual who does not understand the language the barrier is as complete as his counterpart or interpreter wishes to make it."[23] Obviously it was possible for KMAG unit advisors to work successfully with their counterparts without knowing any Korean. They did it. However, advisors who did not know or try to learn the Korean language expressed greater difficulty, more frequent frustrations, and a stronger dislike for their advisory assignment than those who attempted to learn some Korean.[24]

Developing Rapport with Counterpart

A critical task for an advisor was establishing an effective working relationship with his ROKA counterpart based on mutual respect and trust. Without this rapport, no KMAG advisor could perform his mission. Ignored or unheeded, the best of advice was useless. Professional competence was the fundamental requirement that a ROKA commander expected from an advisor. Previous combat experience helped. Many, but not all, regimental and divisional advisors found that advising units one to three levels above their rank and experience was both professionally challenging and rewarding. However, not all counterparts worked well with advisors junior in rank. No matter how good the advisor or how willing the commander, it took time, patience, perseverance, understanding, and luck to develop trust and confidence. Working through interpreters and with limited cultural knowledge, advisors found misunderstandings and confusion the norm. Rapport became the end product of a continuous process of building confidence in each other's honesty and judgment. Requiring hard work to gain, it proved easy to damage. While rapport required professional competence, it rested on the personal relationship between commander and advisor.[25]

For a KMAG advisor to understand his counterpart was not a simple task. As Ridgway observed, "Their unfamiliarity with our ways and our inability to breach the language barrier with consistency, combined with the blundering nature of so many of our dealings with their nation, made cooperation extremely difficult, particularly when the pressure of mortal danger allowed no time for planning or protocol."[26] First, an advisor had to understand his counterpart as a Korean national, then as a ROKA commander, and then as a person. ROKA, founded initially as a National Police constabulary, was formally created in 1949. Some ROKA senior officers had military experience as noncommissioned officers in the Japanese army or with the national police force; others gained experience fighting guerrillas before the war. This meant ROKA officers, regardless of rank, had limited military training and experience. Ridgway observed that a ROKA division commander had the experience level of an average, young US Army captain.[27] Lacking experience and professional knowledge, division and regimental commanders needed advisors. KMAG advisors were older and more experienced than their counterparts, but junior in rank. However, some commanders resented this need; others resented what they perceived as a superior attitude by some American advisors. Getting to know a ROKA commander depended on his willingness to be known and the willingness and ability of his advisor to get to know him.

ROKA was not the US Army. Its capabilities and limitations were far different from the US Army. ROKA commanders were more sensitive to rank and perceived slights. The commander made decisions; no one else. His staff might assist in the execution of decisions, but it did not participate in decisionmaking. Under these conditions, subordinates were reluctant to show initiative, to deviate from the plan, to admit errors, or to submit accurate reports if they reflected poorly on the commander or the unit. Discipline could be harsh, perhaps arbitrary. Commanders seldom asked for advice. Experienced advisors learned to ask timely questions, such as "do you think that it would be a good idea to . . ." or "what are you going to do about . . . ," and to discuss options with the commander without forcing a decision. No commander wanted to be viewed as a puppet to his advisor. The best results came about when the advisor became the key assistant to his counterpart, a true member of the team.[28]

Even when the commander was willing to listen, the advisor found it difficult to ensure he had useful advice to offer. With limited situational awareness, it was easy to come up with inappropriate solutions to actual problems or solutions to problems that did not exist. As one advisor warned, "Often advice is bad because the advisor does not have the whole truth. They only tell you what they want you to know. Advice which is given without the facts or based on inaccurate facts is seldom followed. It shouldn't be. One had to be cautious about such matters."[29] Once good advice was apparent, then the advisor's problem became to determine the best time, the appropriate place, and the most effective manner to offer it.

Over time, the more effective unit advisors were able to establish a good working relationship with their commanders. Some understood that the US way was not always the best or the appropriate way. They came to understand some Korean ways; a few even were able to acknowledge that certain ROKA tactics, techniques, and procedures might be superior to US Army practices. KMAG advisors believed patience and tact were critical personal characteristics. Unfortunately, not every KMAG advisor was up to the task—some lacked ability, others lacked temperament, and others lacked desire. One advisor offered, "People who cannot treat Koreans as equals or cannot recognize certain so-called shortcomings as simply differences in customs, conditions, and language should not be allowed in KMAG." Other traits considered important for an advisor were emotional stability, friendliness, and good humor.[30]

US Army Pressures: Formal and Informal

Throughout the war, KMAG stressed that advisors advised; they did not command. At the same time, EUSA combat commanders demanded effective combat performance from ROKA divisions as well as accurate and timely reports. When ROKA units collapsed, no one was safe, personally or professionally. US combat commanders expected results, not excuses. KMAG told advisors clearly in 1953 that they "stand or fall with [their] counterpart"; advisors were held responsible for their units without command authority. As one advisor made clear, "In an American Corps the Senior Division Advisor *better* feel responsible, for the corps commander certainly considers him so."[31] The preferred advisory tools of discussion, argument, and persuasion were sometimes overcome by events or were inappropriate for these circumstances. KMAG advisors then faced a dilemma. If they threatened or made unfavorable reports to EUSA, they risked their hard-earned rapport and future effectiveness with their counterpart. Yet, failing to report concerns carried grave risks.[32]

An important and unique situation developed in South Korea in July 1950. President Rhee placed ROKA under EUSA command. He agreed to support the EUSA commanders in their fight for South Korea. At times, EUSA commanders relieved ROKA commanders and, once, even

abolished the ROK III Corps. US officials refused to support expansion of ROKA until the Koreans demonstrated their ability to provide effective leadership to their army. This firm US resolve, supported by President Rhee, was felt and believed throughout the ROKA. ROKA commanders knew they ignored advice, particularly in critical tactical situations, at their peril. Yet, even this unprecedented common approach confronted major obstacles and innumerous difficulties, many of which were not surmountable. One KMAG advisor had four different ROKA commanders, all considered unfit for command. The last commander faced court-martial and was exonerated by his ROKA superiors. The KMAG advisor was then removed for ineffectiveness.[33] From the top in this war, both through EUSA and ROKA channels, a similar message was sent: ready or not—perform or else.

Counterpart Observations

Just as advisors had to adjust to their counterparts, counterparts had to adjust to their advisors. Not surprisingly, they did not always agree on things. Many ROKA commanders felt that Americans were "too rude and impatient," that misunderstandings and ignorance arose from differences of customs, and that the language barrier prevented "knowledge of mutual courtesy." When asked, "what was the most important qualification for an advisor," most Korean officers chose a good personality. Although professional competence and combat experience ranked second, the Korean officers emphasized that personality was *"considerably* more important." Willingness to work together in respect and trust was more important to ROKA commanders than competence.[34] A ROKA corps commander emphasized the importance of personality and the willingness to get along with Koreans. He believed that a number of KMAG advisors had been abandoned during the 1950–51 retreat from North Korea because they lacked empathy and respect for their counterparts. ROKA units that considered KMAG advisors "their most prized possession" worked to bring their advisors out safely, and those with little use for their advisors left them to their own devices. He believed that those who had developed rapport with their counterparts were brought out; those without were not.[35]

A Special Study

In the summer of 1953, the US Army authorized the Operations Research Office to conduct a study of the advisory effort in Korea. Based on interviews, surveys, and documents, a draft report was circulated for comment. Published in February 1957 as *The KMAG Advisor: Role and Problems of the Military Advisor in Developing an Indigenous Army for Combat Operations in Korea*, the final report concluded:

1. Each of the assigned duties of KMAG advisors to local national tactical units was necessary and could not be safely reduced without compromising the success of the operation, even though these duties placed a heavy burden on officers serving as advisors. KMAG advisors were usually confronted with problems and responsibilities normally encountered by officers two ranks above their own.

2. Advisory duty in a tactical unit of a local national army, particularly under combat conditions, is exceedingly difficult and frequently frustrating and personnel selected for such duty must be temperamentally and physically able to withstand these stresses, in addition to being professionally competent. Qualities needed include tact, patience, emotional stability, self-sufficiency, self-discipline, and—in tactical units—command and combat experience if possible.

3. The size of KMAG tactical detachments as provided in Korea during combat operations was at minimum practical levels, considering the multiple mission[s] assigned. The pressure of the advisory job was most acute on the regimental advisor in infantry units during the shooting phase of the war.

4. Living constantly with local national army tactical or isolated units, support regiments removed from personal association with other US personnel had adverse effects on advisors' morale and efficiency. KMAG advisors in combat units needed the relaxation offered by periodic social contacts and off-duty companionship with other US personnel, and more frequent R&R than personnel serving with US units.

5. The KMAG advisor had to recognize that certain practices of a local national group, such as the "welfare fund," were deeply rooted in the national culture, and that the advisor's responsibility was to see that these practices did not jeopardize the military effectiveness of the unit.

6. In tactical units the success of the advisor's mission, his personal safety, and sometimes his life, depended on his relation with his ROKA counterpart. For a KMAG advisor to work effectively with his ROKA counterpart it was important that he:

a. Establish rapport based on both mutual confidence and respect for ability, professional competence, and experience and mutual regard and consideration for integrity and personality.

b. Practice military courtesy and protocol appropriate to the counterpart's rank and the advisor's level of operation as a member of the counterpart's personal staff.

c. Maintain close and constant association with his counterpart during working hours, including visits to the field, and be available to observe and advise on all matters that arose.

d. Check and inspect closely every day the execution of the counterpart's orders and the performance of subordinates and units in the command.

e. Initiate advice—in private—to the counterpart on all matters requiring attention, with particular attention to premeditated problems and plans, decisions on current matters, and follow-up of orders or supervision of subordinates.

7. When a really important issue was involved and the counterpart would not voluntarily act in accord with the advisor's proposal, the advisor had to assure compliance by bringing pressure on his counterpart.

8. Logistic support of KMAG advisors serving with local national units, particularly in remote or isolated places, was an acute problem that required special attention.

9. Advisors for MAAG-type assignments needed training in the form of a short intensive orientation before being sent to their duty stations.

10. KMAG advisors did not need to know the local language to perform their missions; but some knowledge of the language was an important asset in advisory duties; efforts to learn the language facilitated personal relations.

11. A tour of duty as an advisor in a MAAG is worthwhile professional experience as well as being a highly important military service.[36]

Based on the study and its conclusions, *The KMAG Advisor* report recommended that for US Army future advisory efforts:

1. Selection qualifications for MAAG advisors should be based on:

    a. The officer's professional competence, preferably demonstrated by command experience—including combat command if possible—for advisors to line units.

    b. Special screening of officers and enlisted men for qualities temperament and fortitude to withstand the strenuous psychological and physical demands of advisory duty in tactical units of a local national army, particularly under combat conditions.

    c. Personal characteristics of tact, patience, emotional stability, self-sufficiency, and self-discipline that will enable the officer to work effectively and harmoniously with local national personnel and that will induce a respect and confidence in Americans and the US.

    d. Preference to officers with facility in the local language.

2. Advisors should be given orientation for MAAG-type assignments, preparatory to entering on such duty, and be explicitly briefed on: their advisory duties and responsibilities; the structure, organization, and the known strengths and weaknesses of the local national army; and the culture and customs of the local nationals and methods of working with them. Language study should be encouraged and facilitated by short intensive courses and/or on a self-study basis, unless more thorough preduty language courses are required at the option of the chief of the MAAG involved.

3. During combat operations and during the development stage of an immature local national army the regimental advisor should be provided with at least an assistant advisor, and also with battalion advisors to operate from the regimental detachment.

4. MAAG advisors assigned to local national units in the field should be grouped together and live in MAAG detachments at regimental or higher headquarters insofar as possible, and advisors assigned to tactical or isolated

units where they are removed from normal daily personal association with other US personnel should be required to spend the equivalent of one 24-hour period per week at a higher MAAG or US detachment.

5. The length of continuous assignment for tactical advisors living with advised units in the field under combat or isolated conditions should be not less than 6 nor more than 9 months, and for advisors living in decentralized MAAG detachments 9 to 18 months.

6. Indigenous interpreters in tactical units should be military personnel of the local national army assigned to the US units, MAAG or otherwise, and under the control of the US officers to whom the interpreters are responsible. This control should include discipline; efficiency rating; recommendations through channels to the local national army for the interpreters' promotions, additional schooling (including that in US schools), and awards; and (at the option of the MAAG chief) messing, billeting, and some supplementary pay in money or kind when needed. In nontactical units civilian interpreters should be authorized, but they should be under corresponding US control and direction.

7. Local national officer-interpreters prior to assignment to US commanders and MAAGs should receive training in the service branch for which they are assigned as interpreters (officer's basic course, branch material).

8. The factors found important for KMAG advisors to work effectively with their ROKA counterparts should be referred to, for the information and guidance of advisors in other MAAGs, particularly in underdeveloped or Asiatic countries.

9. MAAG or military-mission type problems should be included in the curriculums of the Army's principal service schools, with particular emphasis in schools for advanced career officers.[37]

Few of the recommendations were adopted. The war was over. KMAG had performed its job—improved the combat effectiveness of ROKA. In 1953, during this study, KMAG did tighten its procedures for briefing and orienting new advisors. It provided each advisor a copy of a KMAG procedure manual. However, the recommendations on selection and preparatory training of advisors, on the size of advisor teams, on interpreters, and on general training for officers were not acted on.

Summary

During the Korean war, KMAG unit advisors confronted not only the challenges of combat, but also that of working with the ROKA whose language, culture, and ways were often incomprehensible. Chosen for advisory duty by MOS and occasionally by rank, KMAG advisors frequently found themselves working with a counterpart two to three ranks above their rank and advising units larger than the ones they had served in, much less commanded. Advisor teams were small, seldom below regiment, and frequently understrength. Without Korean language skills—working through ROKA interpreters with limited English skills—and without any prior preparation for advisory duty, KMAG advisors attempted to comprehend what was occurring around them; to develop suitable, acceptable, and feasible advise; and to communicate that advice clearly, quickly, and convincingly to their counterparts. At the same time, and despite these handicaps, they were responsible for ensuring that US commanders received accurate and timely situation reports. Through hard work, misunderstandings, mistakes, successes, and working together, ROKA had become a large, combat-experienced, and capable military force by 1953.

Ridgway observed in a 1969 interview, "No Army in modern times was ever subjected to the battle stresses, strains, and losses to which the ROKs were . . . in the beginning of the war."[38] Under the most trying combat conditions, ROKA—initially unprepared, then shattered by the North Korean army, hastily reconstituted in the Pusan perimeter, hurriedly thrown forward into North Korea, smashed by the CCF, rapidly rebuilt, and severely tested again by the CCF all in the first year of the war—hung on, survived, and evolved into an effective combat force. Despite the numerous differences and misunderstandings, Ridgway emphasized the importance of the strong support EUSA commanders received from President Rhee in working toward the common goal of a combat-capable ROKA. Of the KMAG advisors who worked under these adverse combat conditions with hastily formed, ill-trained units led by inexperienced leaders and who tried to establish an effective working relationship with counterparts separated by linguistic, cultural, institutional, and personal barriers, Ridgway added, "officers in an advisory capacity, unit advisors . . . really had a much tougher job than fellows in the regular units, a much tougher job."[39]

Notes

1. Matthew B. Ridgway, "How the Korean Army Improved," interviewed by MAJ Caulfield, October 1969, in *A Systems Analysis View of the Vietnam War: 1965–1975*, vol. 7, *Republic of Vietnam Armed Forces (RVNAF)* (BDM Corporation, April 1980), 52.

2. Alfred H. Hausrath, *The KMAG Advisor: Roles and Problems of the Military Advisor in Developing an Indigenous Army for Combat Operations in Korea* (Chevy Chase, MD: The Johns Hopkins University Operations Research Office, February 1957), 16.

3. Robert K. Sawyer, *Military Advisors in Korea: KMAG in Peace and War* (Washington, DC: US Army Center for Military History, 1962), 34–61.

4. Sawyer, 114–139; for coverage of the war from June–November 1950, see Roy E. Appleman, *South to the Naktong, North to the Yalu* (Washington, DC: US Army Office of the Chief of Military History, 1961).

5. Sawyer, 140–148, footnote 160.

6. Appleman, *South to the Naktong*, provides coverage of this phase of the war; see 673–675 for collapse of ROK II Corps.

7. For this period of the war see Billy C. Mossman, *Ebb and Flow: November 1950–July 1951* (Washington, DC: US Army Center of Military History, 1990); Roy E. Appleman, *Disaster in Korea: The Chinese Confront MacArthur* (College Station, TX: Texas A&M University Press, 1989); and Roy E. Appleman, *Ridgway Duels for Korea* (College Station, TX: Texas A&M University Press, 1990).

8. Mossman covers these actions; see Sawyer, 170–171, for disapproval of ROKA expansion; see Clay Blair, *The Forgotten War: America in Korea, 1950–1953* (New York, NY: Times Books, 1987), 876, for advisor losses.

9. Sawyer, 171–181; for results of training, see Paik Sun Yup, *From Pusan to Panmunjom* (Dulles, VA: Brassey's, 1999), 162.

10. For the last 2 years of the war, see Walter G. Hermes, *Truce Tent and Fighting Front* (Washington, DC: US Army Office of the Chief of Military History, 1966).

11. Hausrath, 95.

12. Sawyer, 58; Appleman, *Ridgway*, 350; Hausrath, 95–96.

13. Sawyer, 58–60; Hausrath, 13–16.

14. T.R. Frehrenbach, *This Kind of War: A Study in Preparedness* (New York, NY: The Macmillan Company, 1963), 508.

15. Hausrath, 26–27, 92–93; Sawyer, 163.

16. Sawyer, 52; Hausrath, 101–104.

17. Sawyer, 57; Hausrath, 14, 37–41.

18. Hausrath, 23.

19. Sawyer, 62; Hausrath, 22.

20. *Draft Copy THE KMAG ADVISOR Circulated for Critical Reading by Selected Reviewers* (Chevy Chase, MD: The Johns Hopkins University Operations Research Office, nd.), 121–122, 128, 144. For American culture, see Loren Baritz, *Backfire: Vietnam—The Myths That Made Us Fight, The Illusions That Helped*

Us Lose, The Legacy That Haunts Us Today (New York, NY: Ballentine Books, 1985).

21. Sawyer, 1, 62–65; Hausrath, 126.
22. Hausrath, 68–71.
23. Ibid., 126.
24. Ibid., 44.
25. Ibid., 48–52.
26. Matthew B. Ridgway, *The Korean War* (New York: NY: Da Capo Press, 1967), 4.
27. Ridgway, "How Korean Army," 47.
28. Sawyer, 65–66; Hausrath, 58–63.
29. Hausrath, 60.
30. Ibid., 28–29.
31. Ibid., 20.
32. Ibid., 82–85.
33. Ibid., 28.
34. Ibid., 28–30.
35. Ibid., 116.
36. Ibid., 2–3.
37. Ibid., 3–4.
38. Ridgway, "How Korean Army," 46.
39. Ibid., 52.

Chapter 2

"A Most Difficult and Sometimes Frustrating Task"[1]

Field Advisors in Vietnam (1961–73)

> You are still the 'heart and soul' of our total commitment to South Vietnam. . . . Your job is a most difficult and sometimes frustrating task. Under any circumstances, the relationship of advisor-to-advised is a testy and tenuous one. Here, that relationship is compounded by daily decisions with life or death consequences, and by communications problems complicated by language difficulties and different national origins. The training of the US military officer is characterized by conditioned traits of decisiveness and aggressiveness. The essence of your relationship with your counterpart is constituted by patience and restraint. As a threshold to development of a meaningful affiliation with your counterpart you must succeed in the reconciliation of these contrasting qualities.[2]
>
> —GEN William C. Westmoreland, Letter to Officer
> Advisors, "US Advisor/Vietnamese Counterpart
> Relations," 1967

Vietnam provided the United States military its longest, largest, and most complex advisory effort. Begun in 1950 when the United States provided logistical support to the French in Indochina, the Military Advisory and Assistance Group, Indochina, became the Military Advisory Assistance Group, Vietnam (MAAG-V) in 1955. Limited by the Geneva Accords of 1954 to 342 personnel, an augmentation of MAAG-V with the 350-personnel Temporary Equipment Recovery Mission (TERM) in 1956 increased its maximum strength to 692.[3] From 1955 to 1960, MAAG-V worked with the government of President Ngo Dinh Diem in building the Republic of Vietnam Armed Forces (RVNAF) into 7 infantry divisions, an airborne brigade, a marine group, 4 armor battalions, a coastal naval force of 10 small ships and 18 amphibious craft, and an air force with a fighter-bomber, 2 transport, and 2 small aircraft observation squadrons. RVNAF focused on its primary mission: defense of the northern border against North Vietnamese invasion.[4] Other nonmilitary US agencies trained and equipped the Civil Guard and the Self-Defense Corps. They handled internal security matters, not MAAG-V. When confronted in 1959–60 with an insurgency instead of an invasion, RVNAF found itself committed to counterinsurgency operations. Given its inherent weaknesses—politicization of its officer corps, rampant corruption, divided command, and ill-trained soldiers in understrength combat units—RVNAF was unprepared despite the 5-year MAAG-V effort.[5]

MAAG-V/MACV During Combat Operations

From 1961 to 1964, the US military effort, both advisory and combat support, increased in response to the deteriorating situation in South Vietnam. MAAG-V advisors, assigned in May 1959 down to regiment for infantry units and to battalion for artillery, armor, and marine units, were permitted in mid-1961 to accompany RVNAF battalion and company units in combat to observe and to offer advice. Advisors were forbidden from direct participation in combat and from participating in operations near international borders. At the Honolulu conference in December 1961, Secretary of Defense Robert S. McNamara authorized MAAG-V to assign an advisor to each province and an advisory team to each operational combat battalion. In addition, he provided US combat support units for the South Vietnamese. Special Forces teams rotated in and out of South Vietnam on temporary duty as mobile training teams (MTT) to train RVNAF units and ranger companies and to work with the Montagnards through the covert CIA-sponsored Civilian Irregular Defense Groups (CIDG) program. At the same time, RVNAF increased its strength, requiring more advisors. To oversee these additional responsibilities and resources, on 8 February 1962 the Military Assistance Command, Vietnam (MACV) was created with General Paul D. Harkins in command. During 1963, RVNAF had grown to 4 corps, 9 divisions, an airborne brigade, a Special Forces group, a marine brigade, 3 separate regiments, 19 separate battalions, and 86 ranger companies. Bigger was not necessarily better. The combat effectiveness of many RVNAF units was not up to fighting the insurgents successfully. By the end of 1963, MACV strength had grown to 16,263 personnel with 3,150 assigned to MAAG-V, of which 1,451 were advisors.[6]

Political instability followed the death of President Diem in November 1963. Not until mid-1965 did the Thieu-Ky military government establish some control over the factions in RVNAF. During this critical period, the situation in South Vietnam worsened. Viet Cong (VC) attacks became frequent and more successful. In the midst of this situation, General William C. Westmoreland assumed command of MACV on 20 June 1964. To improve local security forces, MACV established district advisory teams under each of the 44 province advisors. By the end of the year, the 7th Special Forces Group deployed to South Vietnam to work with the CIDG and to conduct covert operations astride the borders of South Vietnam. By the end of 1964, operational requirements for RVNAF units precluded pacification operations and much-needed retraining.[7] As in Korea, in an attempt to improve the capabilities of RVNAF units, Department of Defense (DOD) asked MACV about providing US cadres for RVNAF battalions. Options considered included US officers and senior noncommissioned officers; US staff officers and technical personnel; and US fire support elements. MACV believed all three options faced the problems of "the language barrier, increased exposure of US personnel, the difficulty of US personnel adapting to ARVN living conditions, and the greatly expanded support requirement that would be generated" and concluded that "US assumption of command was neither feasible nor desirable, owing to the language barrier as well as the probable non-acceptance by the GVN."[8]

From 1965 through 1968, the US military buildup absorbed the attention of MACV. As US combat strength increased from 184,000 in 1965 to 385,000 in 1966 to a maximum of 550,000 in 1968–69, MACV became more an American operational headquarters focused on fighting the North Vietnamese Army (NVA) and main force VC units rather than on fighting the insurgency.[9] For various reasons, Westmoreland decided to accept unity of effort rather than unity of command. RVNAF units did not serve under US Army commanders.[10] MACV and the RVNAF Joint Chiefs of Staff (JCS) developed annual Combined Campaign Plans (CCP) detailing the roles and tasks for both US and RVNAF forces.[11] During this period, the RVNAF focused on pacification. Without any special training for revolutionary development or pacification, RVNAF battalions and their combat advisors dispersed to provide security and conduct operations in populated areas. When this training deficiency became evident in 1967, mobile training teams (MTT) were organized to train all RVNAF maneuver battalions supporting pacification. By 30 September 132 of 144 battalions had received the training.[12] Although the US corps commanders were designated senior advisors to RVNAF corps commanders, they focused on US combat operations and left their advisor duties to their deputies. However, for MACV unit advisors, this meant they now had a US operational chain of command and their counterparts had a separate RVNAF chain of command. Seldom were the two in synch. Some initial efforts at combined operations during this period were generally no more than separate US and RVNAF operations conducted in the vicinity of one another.[13] By the end of 1966, "the buildup eclipsed what had previously been an advisor 'show'" and created additional advisor duties for "law and order, morale and recreation, post exchange, base development, liaison, and visitor and community relations." This caused an "inevitable decrease in the attention to the primary mission" of advising RVNAF and that "liaison was [now] one of the most demanding requirements" placed on advisory teams.[14]

In April 1967, the creation of the Civil Operations and Revolutionary Development Support (CORDS) program under a civilian deputy to the MACV commander brought new priority to the counterinsurgency advisory effort. By combining the 970 advisors of the Office of Civil Operations, who were responsible for revolutionary development and pacification, with the 2,260 MACV district and province advisors, the new CORDS organization had 1,413 military and 1,240 civilian advisors. Not only did province and district advisory teams get more attention, but also a major effort was undertaken to upgrade the regional force (RF) and popular force (PF), formerly known as the Civil Guard and the Self Defense Corps. An early CORDS study discovered that the ratio of US advisors to RF/PF personnel was 1 to 929, but the ratio in RVNAF was 1 to 23. To address this shortfall of 2,243 advisors for RF companies, Mobile Advisory Team (MAT) and Mobile Advisory Logistics Team (MALT) concepts were tested and evaluated. In late 1967, MACV decided to create 354 MATs to work with RF/PF units under province and district advisory teams.[15]

Despite the virtual destruction of the VC and the lack of a national uprising during the 1968 Tet Offensive, the direction of the war changed. Peace talks started, South Vietnam mobilized, and MACV began planning a withdrawal. General Creighton Abrams, who became MACV commander in mid-1968, pushed a "one war" concept. Henceforth, MACV focused on improving RVNAF combat capability and on supporting CORDS pacification efforts. The CCP for 1969 declared, "RVNAF must participate fully with its capabilities in all types of operations . . . to prepare for the time when it must assume the entire responsibility." Emphasizing combined operations, Abrams hoped US commanders and units would model how to do combat operations for RVNAF commanders. For the remainder of the war, most US combat units worked alongside RVNAF, RF/PF, and the newly created, part-time People's Self-Defense Force (PSDF) militia during both combat and pacification operations.[16] Creating 354 MATs to upgrade RF/PF units remained a major CORDS program, but it proved "expensive in terms of . . . experienced infantry officers and NCOs." By the end of 1968, the MACV advisory team had reached 11,596 personnel— 10,544 Army, 615 Marine Corps and Navy, 437 Air Force.[17] This meant that in addition to 11 division equivalents of US combat units in South Vietnam, the "US Army advisor contingent, in terms of officers and senior NCOs, was the equivalent of another seven US Army divisions."[18]

When the Nixon administration announced its Vietnamization policy in 1969, it became clear that the US military effort would decrease and RVNAF would need to improve and to expand to meet security requirements. MACV division advisory teams converted to combat assistance teams (CATs), reducing advisory spaces in the division and changing the advisor's role "from advisory to combat support coordination."[19] The US/RVNAF attacks into Cambodia in April–May 1970 indicated improved RVNAF combat performance, even operating without advisors for a time. MACV increased MAT requirements to 487 teams in 1970 to continue the upgrade of increasing RF/PF and PSDF units. MAT became "the largest advisory element"; however, finding captains to serve as MAT senior advisors proved impossible. By October, less than 80 captains were assigned to MATs.[20]

Although the February 1971 Lam Son 719 operation into Laos raised concerns about the RVNAF capability to conduct major operations, the withdrawal of US combat units and the reduction of US advisors continued. By 30 June, battalion combat assistance teams (BCATs) were gone, except in a couple of units, and regimental combat assistance teams (RCATs) began phasing out on 1 September. During 1971, MATs were reduced from 487 to 66; the remainder scheduled for elimination in early 1972.[21] The US "advisory emphasis had shifted from tactical operations to the function and technical areas and the level of effort in terms of advisors decreased as . . . [the South Vietnamese] gained expertise in various areas."[22] By the end of 1971, only 3,888 tactical advisors remained with RVNAF combat units; many were short-time personnel, fillers from departing US units assigned to complete their 12-month tour.[23]

On 30 March 1972, the NVA launched the Easter Offensive, a phased, three-pronged conventional attack with tank and artillery support. RVNAF divisions, supported by US air power and accompanied by US advisors, fought continuously for several months before forcing the NVA in September to consolidate its limited gains in South Vietnam. Some of the fiercest fighting occurred during the defense of An Loc and the recapture of Quang Tri. At An Loc, MACV advisors played critical roles in coordinating massive US tactical air support, in providing daily situational updates, and in encouraging RVNAF defenders by their continued presence on the ground. Without those US advisors, many believed that the defense of An Loc would have collapsed. Interestingly at An Loc, the RF/PF performance was considered better than that of RVNAF regulars, a direct result of training and better leadership.[24] Although severely tested, RVNAF did not collapse; nevertheless, significant weaknesses were identified in RVNAF and in its leadership. President Nguyen Van Thieu replaced commanders, but the results would not be known until 1975.

In the fall of 1972, two separate studies in Washington, DC, addressed the US advisory effort for 1973. A Joint Chiefs of Staff review recommended cutting the overall advisory effort but continuing battalion advisor teams for the airborne and marine units that bore the brunt of the fighting, reducing the division combat assistance teams (DCATs) from 36 to 15 personnel, and continuing the support of CORDS with advisors at province and district. A separate DOD study recommended a total of 2,500 advisors—380 tactical advisors (10 per division and 50 each for air force and navy), 500 staff advisors (20 per corps, 100 each for air force and navy, 220 for RVNAF Joint General Staff), and 1,620 province or district advisors.[25] None of this mattered. On 28 January 1973, a cease-fire went into effect and the withdrawal of the remaining American military personnel followed within 60 days. After an almost 20-year effort, RVNAF was a battle-tested, well-equipped force of 550,000 regulars and 525,000 territorials. It had an air force and the largest helicopter force in Asia.[26] Yet it still had chronic problems—weak leadership, corruption, unwillingness or inability to reform, weak support from the people, and lack of confidence. When the North Vietnamese attacked again in 1975, RVNAF proved inadequate for the task.

Field Advisors

Just as with Korea, the following focuses on those who performed daily, face-to-face advisory duty. For Vietnam, this includes MACV combat unit advisors with infantry regiments or marine brigades with infantry, airborne, ranger, marine, armor, or artillery battalions and CORDS pacification advisors on province or district teams or with MATs.[27] This survey does not include Special Forces personnel working in the CIDG program, but it does include those on province or district teams. By 1970, MACV field advisory strength reached a high of 14,332. This included 2,976 combat unit advisors—regiment, battalion (infantry, marine, airborne, ranger), and other teams; 5,685 CORDS advisors—province and district teams; and 2,305 advisors on MATs. If the 1968 advisor strength was 7 division equivalents of officers and senior noncommissioned officers, the number was 8.65 division equivalents by 1970.[28] An advisory effort of this scale generated a manpower requirement that was seldom fully met. Throughout the advisory effort, teams were often understrength and filled with personnel who did not meet the stated requirements for the position.

Advisory Structure

Prior to 1961, MACV division advisory teams were authorized an infantry colonel as the division advisor, an infantry major and two noncommissioned officers for each of the three infantry regiments—no advisors were assigned to the three infantry battalions of four companies each, and a field artillery major and three noncommissioned officers for the artillery regiment.[29] To address an increasing insurgent threat, the Secretary of

Defense approved battalion advisory teams in December 1961. Authorized an infantry captain and a noncommissioned officer, many advisory teams consisted of volunteers. In any event, the immediate requirement to create battalion advisory teams for each RVNAF infantry battalion and the increase in RVNAF battalions meant that captains did not lead all teams. Frequently, a young US Army first lieutenant with 2 years service and no combat experience became the advisor to a commander twice his age who had 25 years of combat experience.[30] In 1962, two USMC captains were authorized for each RVNAF marine battalion.[31] Expansion of the RVNAF marine corps later led to the fielding of brigade and division advisory teams and the upgrade of battalion advisors to a major and captain. By 1964, infantry battalion advisory team authorizations increased to five personnel—a captain senior advisor, a first lieutenant assistant advisor, and a communications and two light weapons noncommissioned officers.[32] Artillery battalion advisory teams and armor units were authorized a branch-qualified captain and a senior noncommissioned officer.[33] From 1965 to 1969, the unit field advisory team authorizations remained constant: regiment with 3 personnel, infantry battalion with 5, division with 52, and corps with 143. Even so, at times personnel shortages reduced actual team strength by as much as 50 percent.[34] In 1969, divisional advisory teams converted to CATs. Although the DCAT was reduced in strength, the RCAT and BCAT remained unchanged.[35] During the drawdown, BCATs were eliminated by 30 June 1971 except for airborne, marine, and the newly-raised infantry battalions. RCATs began phasing out on 1 September 1971.

Advising combat units was considered normal advisory duty. However, counterinsurgency generated a demand for civil-military advisors. When battalion advisor teams were approved in December 1961, a province senior advisor (PSA) was provided to 39 of the 44 province chiefs. By 1966, there were 44 province advisory teams.[36] Normally, province chiefs were military men responsible for all governmental and pacification programs in each province. In 1964, the MACV province advisory team was authorized 16 personnel: a lieutenant colonel province senior advisor; a major assistant province advisor; four captains each responsible for advising intelligence, operations and training, psychological warfare and civil affairs, or RF/PF units; a USAF captain air liaison officer; a lieutenant assistant RF/PF advisor; three senior noncommissioned officers each responsible for advising on intelligence, light weapons, or infantry; and five enlisted men for communications, medical, and administrative functions.[37] Just as with combat advisory teams, shortages of qualified personnel were not unusual. In 1967, the creation of CORDS consolidated all civil-military advisory teams for the first time under MACV. On 28 May,

Ambassador Robert Komer became Westmoreland's deputy for CORDS. Shortly thereafter, DOD developed a program to upgrade the quality of military personnel serving in PSA or deputy PSA positions. At that time, civilians held PSA positions in 21 provinces with military PSA in the remaining 23. Because of the mix of civil-military activities, if a PSA was military, then the deputy was civilian; if civilian, the deputy was military.[38] In November 1968, PSA positions were upgraded to colonel, but by January 1970, only one-third were filled by colonels.[39] By 1970, province advisory teams generally consisted of at least 25 military and 19 civilian personnel.[40]

Each province was divided into districts governed by a RVNAF officer. In June 1964, 13 two-man teams of a captain and a senior noncommissioned officer were assigned to districts in seven important provinces. Although this test was inconclusive, MACV decided to create 113 five-man teams—a captain district senior advisor, a first lieutenant assistance district advisor, two senior noncommissioned officers, and a medic. Because each district varied significantly from the others, tailoring district advisory teams was permitted. Of the 181 teams authorized at the end of 1965, 169 were deployed—133 by MACV and 36 by Special Forces personnel.[41] Each of the 236 districts had advisory teams by the end of 1967.[42] By 1970, eight-man military advisory teams, authorized a major as the commander, worked with South Vietnamese district chiefs.[43]

Local security was the primary task of RF/PF units; however, no advisors worked with them until 1968. After preliminary testing, Westmoreland directed in October 1967 that 354 five-man MATs be raised to upgrade RF/PF capabilities. MATs, composed of infantry combat veterans, were authorized a captain team leader, a first lieutenant assistant, and three noncommissioned officers—a light infantry weapons advisor, a heavy weapons advisor, and a medical specialist. Two RVNAF personnel, a lieutenant and an interpreter, joined each MAT.[44] At the end of 1968, by using in-country personnel and creating a MAT advisory course in Vietnam, MACV met its goal of 354 MATs. In February 1970, the program expanded to 487 teams, but it proved difficult to provide experienced personnel.[45] At the end of 1971, only 66 MATs existed, and they were deactivated in 1972.

Advisor Roles

For combat units, prior to 1961 and just as the KMAG, the MAAG-V field advisors focused on training and staff work at division and regiment. As a history of the early advisory effort emphasized, "Not only were most American military advisors unfamiliar with the society, culture and language of South Vietnam, but the advisory role itself was unfamiliar."

According to a senior advisor, the advisory role was "entirely new and challenging to most American soldiers . . . [who] spent most of their lives giving and executing orders. As advisors to South Vietnamese counterparts, they neither give nor take orders; they have a much less positive role—that of giving advice, providing guidance and exerting influence."[46]

Accompanying RVNAF infantry battalions conducting counterinsurgency operations in 1962, MACV unit advisors were to advise and assist "on all matters pertaining to Operations, Tactics and Technique Doctrine, Training, Logistics, and Administration . . . [conduct] inspections to ensure . . . equipment is properly used, secured, and maintained . . . [and provide] information on status of Training and Logistics, on operations, and on major problems."[47] (See Appendix C, "Role of the Individual," and Appendix D, Advisor "Do's and Don'ts," for additional guidance provided to advisors at that time.) The primary focus in the field became tactical advice and coordination of US combat support, fixed and rotary-wing close air support (CAS), and helicopter support for movement, resupply, and medical evacuation. Initial difficulties arose from inexperience with helicopter operations, from incompatibility of equipment, and from language problems.[48] An advisor from this period said he had three roles: a US Army officer following orders and supervising US subordinates, a member of a RVNAF unit sharing its experiences and bonding with his counterpart, and a mediator interpreting and communicating between his counterpart and his US superiors.[49]

From 1965 to 1968, US combat units assumed the major burden of combat operations against NVA/VC units. RVNAF units supported pacification and local security efforts. For MACV unit advisors, life became more, not less, complicated. In March 1965, MACV noted that the tactical advisory effort had evolved from training to tactical advice to combat support. Advisor duties had increased to include "coordinating both artillery and helicopter and fixed-wing air support; acting as a conduit for intelligence; developing supply and support programs; improving communications between combat units and area commands (province and districts); and providing special assistance in such areas as psychological warfare, civic action, and medical aid." MACV considered and rejected redesignating advisor teams as "combat support teams."[50] By the end of 1966, the US troop buildup had introduced additional advisor requirements "in the fields of law and order, morale and recreation, post exchange, base development, liaison, and visitor and community relations." As advisors performed these additional duties, there was "the inevitable decrease in the attention to the primary mission." Twenty-four-hour-a-day liaison had become a "most demanding" requirement for advisory teams.[51]

From 1968 to 1972, RVNAF assumed a greater role in combat operations, but the advisor's role did not significantly change. A USMC advisor, one of only two assigned to his battalion, said his main job was "to be liaison to any American forces, no matter what they are . . . to be sure that the Americans know that . . . Vietnamese . . . are operating in that area, so we don't have . . . friendly fire." He hinted at the difficulty and the frustration in trying to coordinate combat support when planning consisted of "We go now."[52] During this period, MACV advisors described their primary field roles as not advising but "the continued coordinating with US units operating in the vicinity, obtaining and controlling light fire teams (heavily armed helicopters), US Air Force tactical airstrikes, US artillery support and helicopters for general use"; and as an "expediter" to get their counterparts to do what they knew to do, but put off doing.[53] In 1969, unit advisor teams became CATs. Not only did the name change, but the mission changed "from advising to combat support coordination."[54] Until the US withdrawal in 1973, the coordination of US combat support assets and liaison with adjacent US and RVNAF units continued as the major duties of MACV unit advisors.

For the pacification effort, before 1964 MACV advisors assisted province chiefs, but no one worked at district level. That year, advisory teams were assigned to district chiefs and the province advisory teams expanded in size. At both levels, the teams worked security, political, economic, and social programs to improve the governance of the population and to increase its support for the GVN. The varied skills required, many beyond normal military expertise, made working at province or district demanding for soldiers, even without the cultural and language challenges. In 1965, when pacification focused largely on security, province and district advisors had to "work closely with not only their counterparts but also a myriad of higher, lower, and adjacent elements, all of which had to come together before anything substantial could be done."[55] The broad range of duties for senior advisors included implementing pacification plans; advising their GVN chief on security, civil affairs, psychological warfare, RF/PF issues, and RVNAF cooperation; coordinating all US and RVNAF operations within the province/district; and serving on a joint committee with the GVN province/district chief and the Department of State US Operations Mission (USOM) representative to oversee all civic action programs.[56] A typical PSA often supervised and coordinated the efforts of up to 50 US military officers and 100 soldiers, 20 American and 50 local civilians, and US and third country contract employees in working with his province chief on security and pacification programs that involved RVNAF and US military units, US Agency for International Development (USAID), USIA, and non-governmental organizations (NGO).[57] District advisors faced the same myriad of tasks, only on a smaller scale. Combat unit advisor duty appeared rather simple and straightforward when compared to province and district advisor duty.

In 1967, CORDS assumed responsibility for province and district advisory teams. By providing additional resources, developing better selection programs for senior advisors, and creating the MAT concept to upgrade RF/PF units, CORDS, in conjunction with GVN programs, brought improvement to the pacification programs. Created in 1968, the primary MAT mission was "to advise and instruct" RF companies, PF platoons, and RF/PF group headquarters on defensive techniques—"field fortifications, barrier systems, request and adjustment of indirect fires"; and on small unit tactics—"night operations, ambushes, patrols, weapons employment, emergency medical care, and other topics." Its secondary mission was "to advise and assist" RF/PF headquarters and units in improving administrative and logistic procedures—"personnel accounting and record keeping, awards, promotions, morale and welfare, supply support, maintenance, field sanitation, and hygiene." An additional MAT mission was "to provide a liaison capability with nearby US military forces."[58] By 1971, the MAT program had been expanded to include PSDF units and its mission modified "to advise and assist RF/PF leaders in improving RF/PF and PSDF effectiveness," "to advise and assist RF/PF leaders to improve RF/PF administrative, and RF/PF and PSDF supply procedures," "to provide liaison between the advised unit" and US forces, and "to assist PF/PSDF and other local groups and officials in the development and rehearsal of village/hamlet defense plans."[59] Each district had at least one MAT; most had more.

Advisor Selection and Tour Length

Before the increased involvement in South Vietnam in the early 1960s, MAAG-V duty was not a high priority. No particular selection criteria were required except rank, MOS, and vulnerability to an overseas tour. The underlying selection principle was that "generalists rather than specialists were best suited to the complicated task of advising a foreign army on a wide range of activities." Since most officers were not specialists, this "tended to encourage the attitude that any Army officer was . . . qualified to serve as an advisor."[60] Officially a 1-year unaccompanied tour was the standard for almost all advisors; most served only 11 months. Separated from their families, most advisors considered MAAG-V duty a "hardship tour," something to be endured before the next assignment. The short tour provided little incentive to tackle difficult cultural and language barriers, much less a long-term approach to improving RVNAF. An advisor noted that a RVNAF division commander would get an advisor "for 11 months, and then he'd get a new one. The new one would have to start from . . . zero . . . again. [The commander] had heard everything before, and he knew that the advisor didn't understand the language and that the advisor couldn't be everywhere all the time to see what was going on. . . . He knew all about how to handle advisors."[61] Even before major involvement in combat operations, the 1-year tour had affects on counterpart perceptions and behavior, as well as advisor attitudes.

Combat arms volunteers eager to serve in a combat environment filled most of the two-man MACV battalion advisory teams in 1962–63. Teams generally had an infantry officer, but sometimes a first lieutenant served instead of a captain. MACV advisory duty criteria were generally the same as those for promotion—attendance at appropriate military schools, successful command and/or service with US tactical units, and not having been passed over for promotion. In 1964, when battalion advisors were expanded to five-man teams, when five-man district advisory teams were fielded, and when RVNAF increased its number of battalions, the demand for qualified officers and noncommissioned officers—rank and MOS—exceeded supply. By then, "the importance of military experience in advisory postings at the lower levels had become irrelevant."[62] Not only did many advisors not have the appropriate rank for their position, many did not have the appropriate military experience or MOS. To compound the problem, the 1-year tour in South Vietnam normally meant 6 months advising a combat unit and the rest of the tour serving on a staff or in a different advisory assignment. Advisory duty remained desirable until the buildup of US combat units in 1965–66. By 1966, emphasis shifted from sending the best personnel to the advisory effort to sending them to US combat units. On 21 November 1966, Westmoreland wrote advisors that their work was "difficult and often frustrating" and that "the finest officers and NCOs are made available for assignment to MACV as advisors."[63] However, Westmoreland earlier made clear to his commanders that "the number one priority in importance in this theater of war is the quality of [the] commanding officer of US units."[64] If a shortage existed, personnel were to be transferred from the advisory effort to US combat units. Despite its initial attraction, advisory duty became something to avoid.

If combat unit advisory duty became unattractive, working on a province or district advisory team, a much more complex task outside most soldier's comfort zone, carried even less appeal. MACV policy after October 1966 directed that province and district advisors would serve a full 12-month tour in their positions. Even exceptions made only for "compelling reasons" required approval of the MACV commander.[65] By 1967, MACV recognized that the PSA or his deputy, if the PSA was a civilian, required highly-qualified personnel. On 29 November the Chief of Staff of the Army (CSA) approved a program that provided "substantial incentives to them and their families" for outstanding officers that met these qualifications: combat arms lieutenant colonel with combat and Vietnam service, former battalion commander, outstanding record, and ability to speak or aptitude to learn the Vietnamese language. Officers who accepted would receive special training followed by an 18-month tour in Vietnam. Incentives for volunteers included a personal letter of invitation to the program from the CSA; promotion preference; location preference in South Vietnam; preference for next assignment; special accommodations for family either stateside, Hawaii, or Guam; a 2-week leave with family in Hawaii after 12 months; and a 30-day leave.[66] Yet this program—full of personal and professional incentives and personally supported by the CSA—challenging the best qualified officers to become province senior advisors, received only a 35 percent acceptance rate from the initial group of letters. Two out of three declined. By December 1969, the acceptance rate reached 41.4 percent.[67]

From 1964 to 1970, district senior advisors received no special selection or consideration until April 1970 when the Department of the Army created a district senior advisor (DSA) program. Similar in concept and incentives to the PSA program, this program was not optional. Officers could volunteer for consideration, but unlike the PSA program, DSA selectees were directed into the program. Officers were selected and assigned without their input or approval. Qualifications for the DSA program were rank of major; combat arms—infantry, armor, field artillery, air defense artillery, or engineer; CGSC graduate; prior Vietnam service; suitable personality and temperament; ability or aptitude to speak Vietnamese; and prior company command. Additional incentives for a 12-month tour included a $50 monthly special pay allowance and consideration for advanced civil schooling. An 18-month tour included similar incentives to the PSA program plus guaranteed secondary zone consideration for promotion, no involuntary unaccompanied tours for 5 years, and an invitation to join the Military Assistance Officer Program (MAOP).[68] By the time the DSA program was beginning to get off the ground, the reduction of US troop strength brought it to a premature end.

In 1968, the MAT program was created from in-country assets. The decision to upgrade RF/PF units with experienced infantry personnel, officers and noncommissioned officers, required US Army, Vietnam (USARV) to create an advisory school at Di An, to provide instructors directly from

RF/PF duty, and to provide qualified personnel with 4-to-6-months in-country experience for duty with a MAT. After the creation of the 354 teams, MAT positions were filled by experienced personnel, as available, from the continental United States (CONUS) who spent their 12-month tour with a MAT. To reinforce this priority program, MAT leaders and their assistants received command credit. However, in October 1970 only 80 captains were assigned to the 487 MATs.[69] As with all the advisor programs in Vietnam, the demand for qualified personnel exceeded supply.

Advisor Preparation and Training

In the early years of MAAG-V, just as with KMAG, advisors received no special training or preparation. On arrival, they were provided a general orientation. In 1959, DOD contracted the Military Assistance Institute (MAI) to provide courses on different countries. A Vietnam course focused on general information on the country, its population, and customs. In March 1962, the Kennedy administration issued National Security Action Memorandum 131 requiring counterinsurgency instruction for military and civilian personnel—DOS, USAID, USIA, CIA—and special preparation for service in underdeveloped areas.[70]

With the expansion of the advisory effort in South Vietnam by the creation of two-man teams for each RVNAF infantry battalion, the Military Assistance Training Advisory (MATA) course was created at Fort Bragg as part of the Special Warfare Center in February 1962. Initially a 4-week course designed for four types of students—infantry officers and non-commissioned officers advising infantry battalions or paramilitary units, field artillery officers and noncommissioned officers advising field artillery units, small arms repairmen, and field radio repairmen. The course expanded after three classes to 6 weeks. The initial 4-week program of instruction (POI) devoted 136 hours to academic subjects: area studies, counterinsurgency, weapons, communications, and demolitions.[71] The April 1962 6-week POI had 217 academic hours for infantry battalion advisors—25 hours area study, 46 hours Vietnamese language, 57 hours counterinsurgency operations, 8 hours communications, 12 hours weapons, 8 hours demolitions, 22 hours physical conditioning, and 39 hours general subjects such as land navigation, first aid, and night operations.[72] The MATA course revised its POI frequently to meet the changing needs of advisors in Vietnam, and language training increased to 50 percent of the course.[73]

The purpose of the MATA course was to provide "a working knowledge of the duties of a military assistance training advisor in counterinsurgency operations."[74] It was never intended to make experts in 6 weeks. However, it was intended to introduce advisors to the essential things they needed to be familiar with for advisory duty. Quickly, the focus of the MATA course became a familiarization with the Vietnamese culture and language and a general knowledge of advisor duties, responsibilities, and techniques—not technical or MOS skills. Students were branch qualified. MATA military training—weapons, communications, demolitions, counterinsurgency operations—focused on the techniques and equipment unique to advisors in Vietnam and comprised less than 50 percent of the course. To keep the POI current and to establish credibility, many previous advisors—US Army and USMC officers and noncommissioned officers—returned as MATA course instructors.[75] Students found themselves challenged both by the amount of language training and by a "big library [of required and recommended readings] in each of . . . [their] rooms."[76] A *MATA Handbook for Vietnam* was developed for students, both as a reference for their technical training and as a source of suggestions for advisors (see Appendix E, "Tips to Advisors"). In October 1967, the US Army published its first and only field manual for advisors—FM 31-73, *Advisor Handbook for Stability Operations*. Developed at the Special Warfare Center, this manual provided the advisor "a ready reference on doctrine and techniques which are employed most frequently in stability operations" and included suggestions for advisors on developing counterpart relationships (see Appendix F, "Counterpart Relationship").[77] As with most military instruction, the MATA course received mixed reviews from students, based on varied expectations and abilities. But many would agree with a USMC first lieutenant MATA student and a later instructor who said, "I left the MATA course knowing that this was going to be a very different experience . . . [and] prepared . . . not to expect everything I would face, but to expect that I was going to be immersed in a very different culture and adapting to that culture and understanding it was going to be complicated."[78]

Vietnamese proved a difficult language to learn. At most, students took away some conversational and military phrases from their MATA instruction. Having native Vietnamese speakers enhanced the introductory language training and put a personal face on Vietnam by exposing students to cultural issues and introducing the more adventuresome to the cuisine.[79] One officer recalled, "Fifty percent of your training was Vietnamese language, small classes, Vietnamese instructors and no more than 8, 10, 12 to a class so you took 4 hours of language a day and you had the tapes and you worked with the tapes at night, then you had 4 hours where you got cultural studies, tactics, operations, field experience, radios, communications—it was a good course."[80] Another said, "In general, you will spend 4 hours a day learning how to be an advisor and another 4 attempting to learn the Vietnamese language. The language training is probably the most important part of the course."[81] If 50 percent of a 6-week MATA course seemed excessive, the 12-week MATA POI for Laos placed even greater emphasis on language training with 300 hours of language instruction out of 420 total hours—71 percent of the course.[82] Having some knowledge of the language was considered important both for the advisor and for his counterpart beyond basic communications. It indicated a willingness by the advisor to try to understand the Vietnamese; it suggested an interest deep enough to learn something about Vietnam. Those going to battalion, province, or district advisory teams received priority for additional language training varying from 8 to 12 weeks at the Defense Language Institute.[83] Even the best students expected to continue to build on their basic language skills once in South Vietnam.

By 1970, a 3-month Marine Advisor Course was taught at Quantico. Besides refreshing basic infantry skills, over half the course was devoted to total-immersion Vietnamese language training. Unlike the MATA course with emphasis on handy phrases and rote memorization, this instruction focused on grammar and structure. But unlike the MATA course, it lacked Vietnamese instructors. Because Marines who had served on interrogator-translator teams in Vietnam presented the language training, a former student described it as "roughly equivalent to making a photocopy from a photocopy. We were two steps away from the original, and not quite as precise in our tones." He declared the result to be a "third language—a tongue alien to both Vietnamese and English speakers but perfectly clear to us co-van."[84] Most sorted out these differences in South Vietnam. However, this example reinforces that in language and cultural matters, close is often not close enough.

On 10 January 1969, the Special Warfare School was expanded and redesignated the Institute for Military Assistance (IMA). Courses taught were a 6-week MATA (ARVN) course for officers assigned as field tactical and logistical advisors, a 6-week MATA (SR NCO) course for noncommissioned officer field advisors, and a 6-week MATA (CORDS) course for officers assigned as members of district and province advisor teams. The first two were updated versions of previous MATA courses. In fiscal year 1972, a 12-week Military Assistance Security Advisor (MASA) course that included 8 weeks of language training was initiated for military intelligence officer advisors at the Province or District Intelligence Operations and Coordination Center (PIOCC/DIOCC).[85] These latter two courses met the CORDS requirement for specialized training, although late in the process, for all province and district team members, not just the senior advisors.

CORDS province senior advisors (PSA) and district senior advisors (DSA) had the most complex advisory jobs—including security and combat affairs, logistical matters, police affairs, intelligence matters, Chieu Hoi program, refugee programs, psychological operations, Hamlet Development cadre, civilian medical services, civil construction and public works, community development, and civilian governmental programs.[86] Yet they received very little specialized training. With the PSA program in 1967, a special 11-week training program was provided in Washington, DC, at the Vietnamese Training Center of the Department of States' Foreign Service Institute. This was expanded to 33 weeks in 1970 for each PSA and deputy PSA (DPSA). Other than training for the PSA/DPSA, most CORDS advisors attended the MATA or a civil affairs course at Fort Gordon.[87] A 6-week basic and a 42-week extended course designed to prepare military officers and civilians for service with CORDS were opened to district senior advisors in 1969. The 6-week POI of 240 hours devoted 60 hours to country orientation/background, 143 hours to operations of which only 57.5 hours addressed the role of the advisor, and 14 hours for language overview. However, the follow-on 42-week POI consisted of 1,125 hours of Vietnamese language study and 135 hours of advanced studies: Vietnam/South East Asia orientation, rural development, advisor techniques, military subjects of concern to CORDS, and updates on the Vietnam situation.[88] Not until 1969–70 could a DSA attend District Operation Courses at the Foreign Service Institute.

Although most advisors arriving in South Vietnam went directly to their units after a brief orientation, there were two exceptions. After 1964, when district advisory teams were created, district advisors received training in Vietnam from the Office of Civil Operations and then CORDS after 1967. By February 1968, a *Handbook for Military Support of Pacification* was available for district and province team members.[89] After MATs were created in 1968, USARV established an Advisor School at Di An to train MAT members. Former MAT leaders served as instructors. The 125-hour POI focused on general advisory subjects (38 hours), weapons training (29 hours), tactics (21 hours), and language training with native speakers (37 hours).[90] A *rf / pf Advisors Handbook*, published in January 1971, offered suggestions for "Advising the RF/PF" (see appendix G).[91] The course proved effective in getting the initial teams fielded from in-country personnel in 1968 and in getting replacement MAT members ready for their duty with RF/PF and eventually PSDF units afterwards.

Challenges of the Advisory Environment

Prior to 1960, MAAG-V advisors had worked for years with a RVNAF that valued political reliability over military competence. Divided command with overlapping responsibilities encouraged by President Diem, widespread corruption permitted by Diem, and previous combat experience under the French had created a South Vietnamese officer corps where "few . . . shared, or even understood, the American officers' belief in coordination, team-work, loyalty to superiors and subordinates, know-how, and delegation of authority . . . ideas . . . fundamental to US-style military operations."[92] Advisors, "separated by a wide gulf of culture and language," had "only the vaguest idea of the effect" of their advice. Counterparts appeared to be friendly and open to advice, but the "advice seldom produced significant results." Combined with a short-tour length and favorable reports influenced by a "dogged 'Can-Do' attitude" of most advisors who viewed "all faults in the Vietnamese Army as correctable, all failures as temporary," MAAG-V remained either ignorant or powerless to address major RVNAF deficiencies.[93] After 1962, US advisors received training for advisory duty. How well it prepared a field advisor "to befriend and influence his counterpart within the context of the existing South Vietnamese politico-military system" remained to be determined.[94]

Understanding: Culture and Language

Americans and South Vietnamese lived basically in two different worlds, separated by a "linguistic and cultural barrier . . . that was almost impossible for the advisor to breach."[95] This fact remained despite advisor training programs. At best, the training created an awareness of some cultural differences, but understanding, accepting, and respecting cultural differences by both the advisor and his counterpart was the real need. Many advisors saw Vietnam as both an alien and inferior place.[96] They treated the Vietnamese like "undereducated and underprivileged children." The problem became so widespread that MACV forbade use of the phrase "little people."[97] Lack of respect worked both ways. An advisor observed, "Ironically, what ultimately emerged was a situation in which the Americans look down on the Vietnamese, who were at the same time looking down on the Americans." To many Vietnamese, Americans were "arrogant, blundering, clumsy, gullible, and wasteful." Americans liked dogs, but did not respect the elderly or traditional things. They "insulted everything and everyone . . . by boisterous, intemperate conduct and by ungracious displays of wealth." Rarely were "Americans spoken of respectfully."[98] Bringing these two worlds together was at the heart of the advisor and counterpart relationship—a key to developing rapport.

Many advisors, reflecting their American military culture, found it difficult to understand, much less accept, Vietnamese ways. Just as in Korea, the "concept of face" remained critical. Among the more significant differences—shown through behavior, attitudes, or values—were a "preference for indirectness," both in speech and in confronting issues that was viewed by Americans as deviousness or dishonesty; a "relaxed and fatalistic attitude toward time" viewed as laziness or indifference; the "importance of tradition and ritual" viewed as outdated or irrelevant; a "relative indifference to human beings" not directly connected by family viewed as uncaring or inhuman; the "importance of taboos" viewed as superstition or backwardness; and the "most common criteria of the good life" viewed as alien.[99] Americans saw themselves in Vietnam to change the way things were done, caring little for tradition or how things were done before. Americans saw things as good or bad, black or white, caring little for reconciling differences or contradictions. Americans attacked problems quickly and directly; Vietnamese approached problems slowly and indirectly. Americans analyzed a problem, made a decision, and then developed a plan; Vietnamese started with a decision and worked backwards deciding on what and how to do things. American leadership was about the duties and responsibilities of the position, the institution; Vietnamese leadership was about the commander, the man.[100] It took time, even for the best advisors, to work their way through this cultural labyrinth. In the meantime, most discovered that "things were not always as they appeared to be . . . [many] had formed opinions too quickly . . . one of the most common, most serious mistakes a 'new guy' was capable of making."[101] The advisor had to remember that cultural differences were real and "that he must take them always into account. This seems like gratuitous advice . . . [it was often] ignored by the complacent advisor who has his counterpart 'all figured out'—by American standards—and then is astounded when his counterpart does something 'entirely out of character.'"[102]

Field advisors learned not to judge Vietnamese reactions by appearances of agreement or by smiles on their faces. A 1962–63 battalion advisor found that his counterpart "would do it in the typical oriental fashion and say, 'Yes, you are right, we will do that,' and then he would do it his own way . . . rather typical in all of Vietnam in those days. They would tell you what you wanted to hear and do what they thought had to be done."[103] As with most nationalities, "an ingrained dislike of foreigners was central to the Vietnamese outlook. . . . Vietnamese xenophobia was very real. Foreigners . . . often overlooked this because the 'Vietnamese way' dictated that such feeling be concealed. The polite smile and the seemingly obsequious behavior of many Vietnamese was a mask that often concealed contempt for the foreigner." A RVNAF officer reminded an intelligence advisor, "you can't help it if you're American, but you should always remember that very few of our people are capable of genuine positive feelings toward you. You must assume that you are not wholly liked and trusted, and not be deceived by the Asian smile."[104]

Lack of understanding worked both ways. Cultural and linguistic differences, combined with institutional, professional, and personal differences, made "it . . . often difficult for an advisor and his counterpart to understand one another. What one viewed as a reasonable approach to a problem was often viewed as inane by the other. Other than making a sincere effort to understand one another's views, little could be done to close this cultural gap."[105] As one unit advisor observed, "We did not understand what was going on in Vietnam. We were in a foreign land among people of a different culture and mindset. . . . The information sent across the cultural divide was not the information received. There was a disconnect. One thing was said and another thing was heard. One thing was meant and another thing was understood. . . . Meaning, intent, and truth were lost in translation."[106] More than one advisor realized "slowly . . . how little I really did understand. . . . [with experience] I was beginning to have a deeper appreciation of the questions, and that was a beginning."[107]

Despite some familiarization with the Vietnamese language, few advisors developed what they considered adequate proficiency. Training was too short; the language was too difficult. A district advisor declared that language skills are "the single most important prerequisite for success" and that a "raw fact is that there can be no more advisors than there are people able to communicate," either advisor or interpreter.[108] Another added, "The first step for an advisor is to determine a desirable solution to a military problem. The second is to communicate that solution accurately, completely (and persuasively) to his counterpart. Obviously command of the native language is highly useful here."[109] Situational awareness, not to mention understanding, could be jeopardized. "Few advisors mastered the language, and we were often hopelessly baffled and frequently misunderstood," said a 1962 battalion advisor who continued, "not a day went by that I did not find certain situations confusing. I was never entirely certain of what I was being told or what was happening. Most time, being in the dark was of no consequence. But it was frustrating and at times embarrassing. . . . Advisors did not really know the Vietnamese language, and for the most part Vietnamese did not know English. We were strangers, one to the other."[110] An early district advisor with only one team member found language "the biggest stumbling block. . . . I couldn't speak Vietnamese and no one there could speak English . . . the first couple of months, the only thing we did was all with hand signals until we got an interpreter. But even working through interpreters is still difficult."[111]

Hand signal communication was of limited value for skills that could be mimicked, but of almost no practical use to a district advisory team.

Many RVNAF officers, particularly at the higher commands, had some English language training, but few were fluent. Since advisors and counterparts were unable to communicate effectively, Vietnamese interpreters attempted to fill the gap. When available, RVNAF interpreters of varying ability were assigned to combat units and to district and province chiefs. One advisor found that "Interpreters, although useful, have many drawbacks... they introduce inevitable inaccuracies into conversations. . . . [and] discourage the frank exchange of views . . . permitted by a private talk between a counterpart and his advisor." He further recommended using written suggestions because they reinforced the advisor's suggestion and might clear up any misunderstanding, because they provided a memorandum for the counterpart to consider if he desired to delay decision, and because they provided a formality and written authority to the advice.[112] A district advisor in 1968 found "they tell you in school to try and build rapport, whatever that means and so . . . we tried to work on that. I spent an awful lot of time with him [counterpart] . . . but I didn't understand a lot of what was going on. I had an interpreter with me all the time, but the interpreter knew what was good for him and didn't always interpret everything . . . the interpreter was an ARVN guy, not an American guy."[113] Depending heavily on interpreters, most advisors became "victims of the language barrier. . . not fully aware of what was going on around them. . . . This . . . was a crippling weakness, since few interpreters could or would render faithfully what they heard. . . . Failure to interpret accurately and completely was the rule rather than the exception. Many interpreters simply could not understand their American supervisor's English, but they would not dare to compromise their limitations. . . . The most common result was an incomplete and often inaccurate job of interpreting that was bound to lead to misunderstandings. And sometimes the interpreters' inaccuracies were deliberate."[114] One advisor added frankly that the "inevitable consequences of these limitations was that many Americans were flying blind the entire time they served in Vietnam."[115]

Developing Rapport with Counterpart

Developing mutual trust and confidence, rapport between the advisor and his counterpart, remained the mantra for advisor training. Improving the technical proficiency of the unit and establishing a productive personal relationship with his counterpart received emphasis in training, in publications, and in Vietnam. By befriending his commander, MACV expected the field advisor to "influence his counterpart within the context of the existing South Vietnamese politico-military system."[116] Unlike in Korea, no combined military command was established; where EUSA had commanded and demanded, MACV coordinated and suggested. Where President Rhee, in response to EUSA concerns, had demanded better leadership and improvements in combat performance from ROKA, no South Vietnamese government required RVNAF to respond to MACV concerns. Unlike Korea, advisors and counterparts did not receive the same, strong message through their chains of command. Without a common understanding of what was expected from the advisor-counterpart relationship, MACV advisors found that personal rapport became their primary tool for working with their counterparts. The US field advisor was expected to work effectively with his counterpart, but the RVNAF unit commander or province/district chief felt little need to respond unless he trusted his advisor's advice or unless he needed the combat support or civic action assets his advisor could provide.

Establishing credibility and understanding his counterpart facilitated rapport. Advisors without combat experience or rank equal to their counterpart were often viewed as less than qualified—thus, an insult or loss of face. A district senior advisor noted his district chief "did not like the idea that his counterpart was a captain and not a major" as authorized.[117] Another found that "I couldn't do much advising when I got there . . . the district chief that was there had been there for 8 or 10 years. He knew everything about this district."[118] A Marine advisor noted, "Face is not just about embarrassment; face is about power and perceptions of who has it and who doesn't in any given situation. A combat commander who loses face, especially when his personal and professional competence is at issue, can see his power as an effective leader be undercut severely or evaporate altogether."[119] Not surprisingly, few counterparts initially respected a MACV advisor with no combat experience, less military service, less rank, and perhaps a MOS unrelated to the unit. Even though the counterpart may have had years of combat experience, an advisor noted, "what you finally came down to is . . . [the commander] knew more about fighting guerrilla war than I knew because he had been doing it most of his life . . . [but] we had to coordinate all the helicopter gun ships, and we could get air strikes in, and then we could work with people like USOM and USAID, and so we were a kind of go-between between him and support activities."[120] While the MACV advisor "was quite capable of putting 100 percent of his energy into the war, since his stay was short, but [for the counterpart] it's impossible to put 100 percent effort into a conflict for 20 years. And this is what . . . [most advisors] expected."[121] MACV field advisors, normally the most junior and least experienced, found it took time, already limited by duty assignments and tour lengths, and effort by both sides to develop an understanding of their counterpart.

Understanding the South Vietnamese military's structure, roles, responsibilities, values, and procedures was equally important for advisors in developing rapport. It required that "the advisor must learn to recognize and evaluate the relative role of his counterpart in the Vietnamese military's social structure, his freedom of action and of expression, before he can develop a useful working relationship with him."[122] Leading by example, the old American standby, did not always work because RVNAF and American leadership concepts were not the same. While counterparts "might . . . come to admire the American advisor's enthusiasm and effort, they were not necessarily inspired to follow the example."[123] RVNAF military expectations were not those of the Americans. As a 1970-71 USMC advisor later observed:

> By inclination and training, we *co-vans* [advisors] tended to be systems oriented. We sought to establish habits, procedures, and systems that would enable the Vietnamese Marines to move into new surroundings and face new situations and carry out new missions, despite all the accompanying uncertainties. We wanted our Vietnamese counterparts to step forward and seize new responsibilities with confidence whenever the situation demanded, rather than staying in their own backyards of military expertise. We also sought to ingrain, through training useful patterns of service and operations that would help them withstand the rough shock of war and continue to function effectively. Finally, we wanted them to understand and appreciate all the ways in which they could support each other and strengthen the team.
>
> On the other hand, the Vietnamese, to the extent that they still paid homage to the Cult of the Commander, did not see fitting themselves into systems and patterns as all that important. It was the commander, with his knowledge, courage, good instincts, and good luck, with possibly the Mandate of Heaven thrown in, who was all all-important. Systems were things to be either used or circumvented, for the benefit of the commander or his family, or perhaps his units, in some cases. So it was not realistic to expect Vietnamese commanders to convert instantly to cookbook leadership, no matter how many great ideas we might manage to set down on paper.[124]

Besides the "cult of the commander," important differences included the preference for staff positions with promotion and power versus command with responsibility and risk; the importance of political reliability versus military competence; the focus on the status quo and doing the safe, approved-by-higher-headquarters thing versus independent, bold action; the system of rewards and punishments; and different relationships between commanders and commanded.[125] An advisor could not understand his counterpart nor know what was possible without understanding the institutional constraints under which he labored.

Developing rapport was just the beginning, the prerequisite to advising. The advisor "had to go beyond . . . [a good relationship] by coming to grips with the realities of the situation, developing a firm understanding of the problem and communicating his suggestions to his counterpart."[126] An advisor noted that gaining trust and confidence was "only the prelude to his major objective: inspiring his counterpart to effective action."[127] Experienced advisors learned that the most useful advice addressed problems from the perspective of their counterparts working within the RVNAF military system and within its capabilities. This was particularly the case for province and district advisors where knowledge of local conditions and local expectations was essential to doing effective work. Many a well-meaning advisor engaged in projects that made sense to an American, but did little for the Vietnamese. As an advisor noted, "Clearly, an intimate knowledge of the people—their customs, mores, attitudes, values, taboos—is indispensable in conducting intelligent and effective civic actions. Here, the US advisor can be at a real disadvantage."[128] Another advisor was told, "you insist on doing it [improving our living conditions] as if my people were not Vietnamese but Americans. All the things that are good for you are not good for them. . . . A good life here is not the same as the good life in America. You must first ask yourself what the Vietnamese need and want. We must answer these questions."[129] And there was not a universal answer, for local conditions, regional differences, and economic variables all had to be considered.

Different advisors used different approaches at different times with their counterparts. A 1967 battalion advisor said his RVNAF commander "was a seasoned fighter who never spoke to me in English. In fact, he rarely spoke to me but I could see signs that he had confidence in my ability to function in the capacity he desired."[130] Suggestions were provided in various documents to assist advisors. (See appendixes C through F for a sample of the ideas provided from 1962 to 1971.) However, the basic fact for advisors remained that "criticism, no matter how constructive and well meant, is seldom well received, even under the best of circumstances. . . .

The concept of 'face' . . . came into play. . . . Vietnamese commanders at all levels had an ingrained reluctance to be seen as relying too heavily on the advice of their [advisors] . . . many of us had adopted an indirect approach. . . . The subtle approach generally worked for nice little gems and nuggets of ideas, when there was no requirement for immediate actions or decisionmaking. But how could we continue to use this face-saving (for our counterparts), indirect technique of persuasion when practically everything needed fixing simultaneously and immediately" remained a classic dilemma for advisors.[131]

Most advisors came to realize that "determination, patience, and perseverance were the most important virtues demanded of advisors . . . more important than the ability to face danger with confidence and resolution" when working through the daily frustrations and inevitable misunderstandings.[132] Another essential trait was common sense—not what made sense to an American, but what made sense to an American working in Vietnam with his RVNAF counterpart. As a district advisor noted, "There was no course to take or book to read that would guarantee success. He had to feel his way along, charting his course with great care, hoping to avoid pitfalls along the way. Since no two districts were alike, he could not rely on the experience of his peers . . . his greatest asset—common sense. In the end, it was this uncommon commodity that separated the successful advisor from the failure. If he had this everything was possible; without it, nothing could save him."[133] A former advisor offered that in working with his counterpart, a good advisor "overlooks no opportunity to give deserved praise," "avoids . . . the pernicious practice of criticizing . . . behind his back," "remains unobtrusive . . . directing the spotlight on his counterpart," "voices . . . disagreements in private," "praises the good features of proposals" with which he disagrees, and does not "box in" his "counterpart to the extent that you appear to be forcing him to take action in your favor—especially if that action would be unpopular."[134] Not only wrestling with their counterparts to improve RVNAF operations, US field advisors wrestled with MACV requirements and with their American military background.

US Army Pressures: Formal and Informal

MACV expected field advisors to improve the combat effectiveness of RVNAF units or to implement pacification efforts in provinces and districts. Without a combined commander making demands on both MACV advisors and RVNAF commanders to produce results, advice often was not acted on. Most advisors found themselves combat support or pacification support coordinators and liaison officers to US and allied units rather than tactical or pacification advisors to their counterparts. Appropriate advice supported by hard-earned rapport, rather than leverage or formal system, accounted for whatever effectiveness a field advisor had with his counterpart. The frequent turnover of advisors due to 6-month duty assignments and a 12-month tour meant multiple advisors for RVNAF counterparts, but no continuity to the advisory effort. Each time a new advisor arrived, he started over in developing a working relationship with his counterpart and in developing an understanding of the situation of his unit, province, or district. Of course, MACV required regular reports—initially subjective assessments followed by a more comprehensive system. Not surprisingly, "the dogged 'Can Do' attitude of most officers and noncommissioned officers who tended to see all faults in the army as correctable, all failures as temporary," and the expectation of results contributed to inaccurate and overly optimistic reporting.[135] MACV ensured that field advisors "who became too frustrated with the performance of their counterparts, or those whose reports were too critical," were "quietly but promptly relieved and transferred."[136]

American military advisors tended to be confident, self-assured, and results-oriented. Empathy was not an American military strength. "If only an advisor can place himself in the shoes of his counterpart and truly understand and appreciate the counterpart's problems and frustrations, then he can assist in the alleviation of these problems and frustrations," offered an advisor. "Unfortunately, an advisor frequently arrives on the scene with preconceived ideas and charges full speed ahead without the slightest idea or care about the effect that it has on the counterpart."[137] A district advisor described Americans as "playing catch up all the time" but wanting to "fix everything right now" by being "always in a hurry," and having "to do things."[138] Another district advisor ran into a senior US officer who told him, "We can't trust these guys to do that. You get in there and do it for them!" The advisor responded that "I couldn't believe what I was hearing . . . the American leader, a brigadier general saying exactly the opposite of what the advisory courses are teaching."[139] Such misunderstandings about advisory duty were common. Another noted that the "largest single cause leading to senior advisors being relieved was their inability to maintain a working relationship with the district chief. Rarely were they relieved due to incompetence, laziness, or lack of aggressiveness. Ironically, it was often characteristics such as aggressiveness and a willingness to 'get the job done quickly' that led to their dismissal."[140] At the other extreme was the experience of a 23-year old first lieutenant senior district advisor:

I was determined and eager to do my best. Given free reign

by a do-nothing, but compliant district chief, I began to accept a growing list of duties and responsibilities. . . . In many ways I controlled life and death of thousands of the people. . . . With no one around to give me my true measure, I began to accept my elevated status, and I began to use the powers in my hands as if they were mine by right. Most of the responsibilities were not truly mine, but I knew the district chief would approve anything I did, and if I didn't do it, I had the definite impression that very little would get done. Perhaps it was only youthful American arrogance that made me take those powers that were outside my rightful reach, perhaps it was the almost mystical idealism with which I took on my whole task, but when I had the chance to get something done I by-God took it! Perhaps I was just a high-tone American, but in my dreams I was a cavalier for freedom, I was a warrior for Camelot. Even more that that. I was a Warrior King.[141]

MACV field advisors remained what they were: American military personnel with all of their capabilities and all of their limitations.

Counterpart Observations

The US Army Center of Military History sponsored a series of Indochina monographs written by former South Vietnamese military leaders. The following comments are drawn primarily from two of these studies, both published in 1980: *The U.S. Adviser* and *RVNAF and US Operational Cooperation and Coordination*. The term "advisor" in Vietnam implied the "power behind the throne"; a negative connotation that "saving face" required RVNAF commanders to avoid.[142] Consequently, most advisors were simply *dai dien*, or "representatives" of the US government.[143] Marine advisors were *co-van*, or "trusted friend."[144] The overall goal of the American advisory effort was to organize, train, and equip RVNAF and to develop combat effectiveness appropriate to maintain internal security and to defend against external attack.

In the early 1960s, RVNAF commanders "found American training and warfare methods too inflexible, too mechanical, and not realistically adapted to the Vietnam battlefield. The language barrier and cultural difference . . . formed a wide and seemingly unbridgeable gap. To a certain extent, the Vietnamese were not interested in training and did not think it was necessary . . . they were experienced enough and knew how to fight this kind of war. American tactical advice was something they thought they could do without."[145] Thus, RVNAF battalion commanders found MAAG-V company-grade advisors of limited usefulness when conducting combat operations. The advisors tended to focus on technical and logistical issues rather than combat or tactical matters. However, with the increase of US combat support assets, advisors became more important to their counterparts as providers of combat support.[146] From 1965 to 1968, US combat forces focused on combat operations and RVNAF assumed a secondary role mostly confined to pacification support. Many advisors saw a serious decline during this period as RVNAF units "were seldom given the opportunity to develop their combat effectiveness, bound as they were to the tedious task of pacification support and territorial security responsibility. Boredom and routine . . . eroded their combat skill and spirit to the point that they became almost as passive and as lethargic as the territorial forces."[147] After Tet in 1968, RVNAF resumed combat operations with MACV advisors providing combat support and liaison with adjacent American units. Combined operations with US combat units, designed to increase RVNAF confidence and capability, actually caused many RVNAF personnel to "regard Americans as protectors and providers instead of advisors and comrades-in-arms."[148] After the war, several former RVNAF officers thought, "it is unfortunate that US . . . [personnel] at the top echelons of the structure did not push hard enough for [RVNAF] improvements. . . . The advisory effort should have endeavored first to bring about an effective command, control, and leadership system . . . before trying to improve the combat effectiveness of small units. If this priority had been established, the entire advisory effort would have been more beneficial."[149] At any rate, neither South Vietnamese nor American leaders were willing or able to tackle this problem during the war. What was seemingly clear afterwards was not so at the time.

The role of the advisor was acknowledged as "not an easy one." It was hard to understand, accept, or change dissimilarities between the RVNAF and the US—culture, way of life, and military concepts. It was easy to change technical and procedural issues, but it remained impossible to reconcile these differences quickly. This became a matter of personality, trust, confidence, rapport, and time. Nevertheless, impatient and often inexperienced American advisors tended to deluge their commanders with suggestions, plans, and programs seemingly as fast as they could think of them. "To his counterpart . . . it was not always easy to cope with all of them at the same time, because there were certain things the advisor would fail to recognize as difficult or impossible unless he was a Vietnamese commander."[150] Advisors did not always consider their counterpart's situation nor did they always provide advice pertinent to the situation. It was the quality of the advice, not the quantity, that was important to the commander.

Former RVNAF commanders believed that a lack of advisor language skills and the short-tour length adversely affected the advisory effort. A former senior RVNAF officer stated, "I know of no single instance in which a US advisor effectively discussed professional matters with his counterpart in Vietnamese. The learning and development of a new language seemed to have no appeal for US advisors who must have found it not really worth the effort because of the short tour of duty in Vietnam."[151] By the end of the war, many RVNAF commanders had worked with 20 to 30 advisors, and "every change of advisor disturbed the atmosphere of the unit."[152] Further, the "1-year tour seemed not conducive to . . . extensive preparations . . . other than perfunctory requirements and a brief orientation course prior to field deployment."[153] They believed an advisor only really knew his job at the beginning of a second advisory tour. To become a good tactical advisor required "a certain continuity and stability of effort devoted to a unit" and they recommended 18 months as a minimum tour length, with a preference for 2-year tours.[154]

MACV combat unit advisors, although few in number, were considered "instrumental in the gradual improvement" of RVNAF units. Not only did their counterparts learn "a great deal from them," but advisors served "as a catalyst through which changes and improvements were attained" and provided "the incentive that stimulated and spurred actions" by both the commander and his unit.[155] But there were downsides. Reliance on US combat support assets tended to establish the primacy of the advisor during combat operations. One "consequence of over-reliance on material assets as substitutes for initiative and prowess was a failure to develop the infantryman's capabilities to the full . . . organized and trained by US standards and exposed for a long time to US warfare methods, . . . units inevitably became accustomed to conducting operations with an abundance of supporting material resources . . . [and] when American presence and assistance were no longer available, the morale and combat effectiveness of ARVN units became uncertain."[156] Another consequence was an "overriding influence" of the advisor that "sometimes tended to stifle the . . . commander's own initiative . . . diminish his authority and prestige . . . [and] tarnish the role of the ARVN commander in the eyes of his troops." Sometimes this made the commander "excessively reliant and sometimes totally dependent on his advisor." When this happened, "the commander's initiative, sense of responsibility, and personal authority became seriously affected and in the long run, the advisor's presence had the undesirable effect of reducing the counterpart's chances for asserting and developing his command and leadership abilities."[157]

CORDS pacification advisors received high marks for their work. Province chiefs were senior, combat-experienced military officers who understood the local situation better than his senior advisor, whether military or civilian ever could. But, the American provincial advisory effort brought resources, assets, and an advisory staff much more capable than anything available to the province chief. In contrast to province chiefs, district chiefs were young, enthusiastic, and inexperienced; quite incapable of managing three to six RF companies, 40 PF platoons, and thousands of PSDF personnel, much less the myriad of nonmilitary responsibilities.[158] The advisory team "at the district level was truly beneficial for the pacification program and contributed substantially to the general war effort."[159] At the lowest-level pacification effort, district advisory team members were "a special kind of advisor. Because of the combat and social environment in which he lived and operated and the many and highly diversified problems he had to solve, the district advisor at the end of his tour had truly become a political-military advisor in his own right."[160] Working with RF, PF, and PSDF units for province and district advisory teams, the MAT effort was considered the "most important and outstanding among US contributions . . . [for] the expansion and upgrading of the Regional and Popular Forces which in time made up over one-half of the RVNAF total strength and became as modernized in armament as the regular forces."[161] CORDS advisors were "the main source of stimulation, incentive for better performance and more devotion by all Vietnamese concerned with the pacification program."[162] Despite this outstanding work, suggestions for improvement included, "advisors at the province and district levels should have been required to speak the language . . . because this was the only means of developing an insight into the local problems of pacification and developing the kind of rapport with the local people that was conducive to success. . . . The ability to speak the language . . . was a most effective tool of winning the battle for the 'hearts and minds.'"[163] In addition, working in pacification was the most difficult advisory task, a "dual military-civilian one requiring numerous skills and endurance. . . . For a task as demanding and as people-oriented as pacification, those who were involved should have been carefully prepared for it and should have learned to speak the language and to live among rural natives as well."[164]

From the RVNAF perspective, for an advisor "the keys to success were . . . personal attitude and . . . genuine desire to help his counterpart. Mutual respect and understanding were always required. For without mutual respect, nothing could be achieved and no advisor technique could help."[165] Three techniques were offered for advisors to consider in getting a counterpart to act. First, provide a draft plan to the commander. This was considered the equivalent of asking the commander to act now, and it usually worked. However, since it was the advisor's plan, it undercut the commander and left room for blame if the plan failed. Second, provide a remark or suggestion informally to the commander. This indirect, tactful encouragement could work with many counterparts given time. Still, from the RVNAF perspective, many advisors found this approach too slow and difficult because the advisor did not get any credit. Third, provide the commander a written assessment of a problem with options, but without any recommendation. This was considered most valuable for commanders with staffs—best at division and above, but no lower than regiment.[166] While an advisor's personality, his professional competence, and his techniques and procedures were important, "what really mattered . . . [to the] Vietnamese was a correct attitude, sincerity, and mutual respect."[167] Advisor training in skills and techniques was important, but less important than developing the attitude and understanding necessary to develop a personal working relationship with a foreign counterpart.

RVNAF commanders did not doubt that American advisors did an outstanding job. They acknowledged improvements in combat effectiveness, technical knowledge, and managerial skills. This fact notwithstanding, the one thing this effort "seemed never able to achieve: the inculcation of motivation and effective leadership" was perhaps the most critical to success.[168] One added, "In retrospect, the improvement of military leadership, particularly at the higher levels of the hierarchy, would have been more vital for the purpose of developing combat effectiveness for the RVNAF than another program. At the higher levels, what the advisors sought most to do was establish good rapport with their counterparts rather than pressuring to do the job."[169] Applied to the highest level, this agreed with the assessment of Brigadier General James L. Collins, Jr., that the "rapport approach is dangerous because it lends itself to the acceptance of substandard performance by the advisor. In any future situation where advisors are deployed under hostile conditions, the emphasis should be on getting the job done, not on merely getting along with the individual being advised."[170] At any rate, at the lowest levels, if the emphasis through both chains-of-command to the commander and to the advisor had been to get the job done, rapport remained the best tool.

Special Studies and Other Observations

During and after the Vietnam war, special studies and many after action reports addressed the advisory effort. This section will look at the conclusions and recommendations of several: a 1965 RAND study, a 1968 DA study and the resulting Military Assistance Officer Program (MAOP), a 1972 Senior Advisor's debriefing report, a 1980 US Army-contracted study by BDM, a DA study, and a Center for Military History official history.

RAND Study, 1965

In March 1965, Dr. Gerald C. Hickey published *The American Military Advisor and His Foreign Counterpart: The Case of Vietnam.* The purpose of this RAND study was "to suggest ways in which the relationship could be improved so that Vietnamese military authorities would be more likely than they are at present to understand, accept, and act on American advice."[171] Building on 10 years work on Vietnam, including 4 years in country, Hickey interviewed 320 American advisors at 70 locations during a 10-month period in 1964. At the beginning of the study, he gathered background information from US military schools that trained advisory personnel. In Vietnam, he conducted both informal personal interviews and structured, multiday group discussions organized by the Walter Reed Army Institute of Research. Although not permitted to question the Vietnamese directly, Dr. Hickey, fluent in the language, was able to gain insights from listening to conversations.[172]

At the time of the study in 1964, 30 percent of the US personnel in Vietnam were advisors. Hickey began with the following general observation:

> The greater the advisor's professional competence and his ability to establish rapport with the man he is advising, the more likely is it that the counterpart will accept and act on his advice. One quality without the other will greatly diminish the effectiveness of the American. Professional expertise is a requirement both obvious and easily measurable, and it has not been the crucial problem in the advisor-counterpart relationship. A faculty for effective interaction with a foreign national and the skill necessary to developing and expressing that faculty are much more intangible. They play no part in traditional military pedagogy, and their great importance is perhaps not yet fully understood in all quarters that must concern themselves with the novel requirements of counterinsurgency.[173]

From his interviews and discussion groups, Hickey identified "the major principles governing the advisory role in a country such as Vietnam, and the chief barriers to better understanding and cooperation at present."[174]

This included a general review of problems, particularly those created by language, cultural, and institutional barriers. From his study, Hickey offered concrete recommendations on:

- Selection Criteria.[175]

—Advisory service should be voluntary to ensure strong motivation.

—Whether voluntary or not, careful screening of personnel was needed to test the suitability of candidates based on professional competence and experience, adaptability to foreign cultures, temperamental disposition to work with foreigners, language skills or abilities, and the possibility of "culture fatigue" of fully qualified personnel who were no longer enthusiastic about this work.

- Desirable Emphases in the Training of Advisors.[176]

—"Language being the single most important factor in breaking down cultural barriers, language training far more intensive than at present should be given to all field advisors." Advisors at higher levels, where face-to-face, almost daily contact with counterparts is not critical, required a brief course on the general structure and conceptualization of Vietnamese language and on the proper use of interpreters.

—To prepare advisors for the cultural obstacles they would have to overcome or bypass, training programs "must insist on the importance of respecting the Vietnamese cultural identity" and must emphasize the patterns "most strikingly different from ours." These included indirectness in the language and general discourse; relaxed attitude toward time; importance of traditions; unconcern for those not family or close friends; taboos; attitudes toward hygiene and health; and the "most common criteria for good life."

—Training courses should include instruction on both the "special characteristics of the region" and the country itself—history, government, economics, society, ethnic groups, religious sects, general customs.

—The common distrust of Vietnamese food created "at times a barrier to good feeling and camaraderie between advisor and counterpart." Attempting "to break down the prejudice rather than reinforce it" by emphasizing the excellence of native cuisine and the examples of advisors who suffered no ill effects should be the focus of training courses.

—Instruction on the formal structure of the Vietnamese military should be supplemented with how things happen—what advisors called its "real workings"—the unofficial and unwritten ways things were

done. Topics suggested were the French military legacy, decisionmaking techniques and procedures, position and prerogatives of various ranks, rewards and punishments that affect promotion, and how the counterpart fits into his military hierarchy and the limits imposed on his autonomy.

—"Far greater attention . . . [needed] to be given to all facets of civic action." Basic understanding of civic action—its importance, its principles, its planning, its approaches for working with the Vietnamese—was missing. However, once the basics were understood, the "prerequisites for informed civic actions planning" remained "above all, an awareness of specific local needs and wishes (which may be quite different from what American standards would dictate and can be acquired only through familiarity with the region and consultation with the people and its leaders), and a number of local or economic variables that might make an otherwise attractive innovation undesirable." The most efficient way often proved not to be the most effective way. Creating unemployment by doing things "better" with machinery, using "better or easily available" external resources rather than local materials, and ignoring local taboos and superstitions were offered as examples of how not to do civic action.

—Advisors needed to consider the short-term, long-term, and potential adverse impact of any advice offered. Advisors had to "learn to weigh the merits of an immediate objective against any undesirable side effects it might have" and to keep in mind that "by exploiting a temporary advantage . . . an advisor could permanently alienate a counterpart and thus lose his cooperation in more important ventures."

—Instead of having recent advisors address future advisors at US training sites, 3-day "exit—entrance seminars" in South Vietnam were suggested. Outgoing advisors would address in-coming advisors that were to work in a comparable assignment, thus ensuring accurate information for the new advisors.

—A pilot program similar to the language and cultural training centers missionary societies had established in-country might serve as a useful model for preparing advisors.

● Administrative Considerations.[177]

—Minimize bureaucratic requirements on advisors, particularly paperwork.

—Because it took months to develop an effective working relationship with a counterpart, advisor tour lengths needed to be reviewed and extending the 6-month battalion advisory assignment to 9 months seemed suitable.

—Given both professional and cultural sensitivities, developing mutual respect between an advisor and his counterpart was likely to be quicker when the rank and MOS of both matched.

—Special Forces A-teams should continue to rotate as teams, not be individual fillers.

—Vertical communication between advisors and their superiors needed to facilitate frequent opportunities for discussions of roles, issues, and problems that were required to maintain mutual trust and understanding with one another.

—Sharing ideas between advisors would be enhanced by lateral communication. Periodic gatherings of advisors from the same level to discuss common problems and experiences and to exchange techniques and suggestions were recommended. Attendance by representatives of higher headquarters and capturing the discussions on tape to provide a permanent record was suggested.

—To facilitate the advisory transition and to reduce the time needed for a new advisor to become effective, departing advisors should "draft . . . a brief, informal profile of his counterpart and a record of advice already given and either accepted or rejected. A new advisor who is prepared for the personality of his counterpart, his idiosyncrasies, and his receptivity to advice, and who knows what advice has already been tried, will be spared much of the time-wasting trial-and-error phase of the uninitiated."

—Some advisors thought it was time to "explore the wisdom of terminating some advisory functions and reducing others." Some counterparts, benefiting from numerous advisors, reached a point where they required no assistance, or only sporadic help that could be provided by an advisory pool.

—However wise terminating advisors for counterparts "saturated with advice" seemed, planners were reminded that field advisors also fulfilled the "invaluable function of an American observer" who reported on local conditions and situations.

The Hickey study provided an excellent analysis of the many problems caused by the "generalized selection process, limited training, and the relatively short tours."[178] In addition, it provided practical suggestions. But in 1965, MACV began focusing on the role of US combat forces.

Some of these recommendations were reinforced by a 1965 special report by the Army Staff that took into account the comments of over 300 senior advisors. That study concluded that "the entire advisory system needed to be strengthened by a unified chain of command, greater control over direct and indirect American military support, longer tours, and a comprehensive debriefing and evaluation program for departing advisors."[179] Without a combined US-RVNAF chain of command, by 1969 another RAND study concluded that the emphasis on "harmonious relations"—rapport—created the attitude that "the advisor only exposes his own incapacity when he complains to his own superiors about the stupidity, want of integrity, laziness, ingratitude, or lack of competence of his counterpart." Such opinions "inevitably hampered the effectiveness of the advisor, undermined the veracity of the advisory reporting system, and masked serious faults in South Vietnamese units."[180]

DCSPER-40 Report (1968 DA Study) and the Military Assistance Officer Program (MAOP)

In May 1965, the Secretary of the Army convened the Haines Board to review the Army officer training and education system for the next 10 years. Lieutenant General Ralph E. Haines, Jr. chaired the board of four general officers, four colonels, and two lieutenant colonels that looked both at officer training and for what officers were being trained. The board looked at special career programs that filled vital needs not normally addressed by branch-material assignments. Its recommendation 40 was "that a modified and enlarged Foreign Area Specialist Program, renamed the Foreign Studies Specialist Program, be established to embrace training in language, regions, psychological operations, civil affairs and related subjects . . . [and] that it absorb the Civil Affairs Specialist Program." This would have centralized control of all functional areas normally considered relevant to general political-military missions—CA, PSYOP, etc.—and of the Foreign Area Specialist Program (FASP), an intelligence-focused area and language program. This, in effect would have created "an Army-wide academy of social sciences and area studies, with a very broad training responsibility and task orientation." When the Haines Board Report was submitted in 1966, recommendation 40 was not adopted. [181]

General Harold K. Johnson, CSA, convened a new study group under the Office of the Deputy Chief of Staff for Personnel (ODCSPER) in March 1967 to reconsider the Haines Board recommendation 40. In March 1968, the final report, known as "DCSPER-40," was briefed to the CSA. The final report recommended that the FASP remain a separate program, but that other politico-military skills and knowledge be organized under a new program—initially called the Overseas Security Operations (OSO), it was soon redesignated as the Military Assistance Officer Program (MAOP). The program aimed to address inadequacies of PSYOP and CA by broadening their focus and by developing a more integrated effort, to remedy a lack of command support for civil-military operations (CMO) by creating a G5/S5 staff section, to develop the integrated and coordinated skills required for successful stability operations, to bring together military functions related to advising host nation military forces, and to focus on operational issues separate from the intelligence focus of FASP. On 26 April 1968, the CSA approved a commissioned officer special career program implemented in Army Regulation (AR) 614-134, *Military Assistance Officer Program (MAOP)*.[182] Initial estimates were a requirement for 6,000 positions.[183] The Military Assistance School, Institute of Military Assistance at Fort Bragg began a pilot Military Assistance Officer Command and Staff Course in September 1969.[184] In 1971, the course was extended from 19 to 22 weeks and limited to prospective candidates of MAOP.[185]

AR 614-134 established the policies and procedures for the Military Assistance Officer Program: to develop officers with the "critical skills needed to serve as commanders and advisors and to man key staff positions in the conduct of military activities having social, economic, political, and psychological impact" with a focus on "developing nations and the positive role of indigenous military forces in contributing to national development."[186] MAOP was designed with two parts—key positions for colonels and lieutenant colonels and supporting positions for majors and captains. Prerequisites for the program were a rank of captain to colonel; US citizen; completion of military schooling appropriate to rank; baccalaureate degree or higher, preferably in the social sciences; exceptional performance record; language proficiency of either minimum qualification on the Army Language Aptitude Test (ALAT) or a R3/S3 (foreign language proficiency); express a desire to participate; a minimum of 3 years active service remaining; and not being a participant in another Army special career program.[187] Officers selected alternated between branch material and MAOP assignments with the final objective to "produce fully qualified officers capable of filling both brand and MAOP positions." The education and training program was tailored for each officer. Captains attended the MATA, MASA, Military Assistance Program Advisor (MAPA), CA Officer, Civic Action, District and CA Advisor, or PSYOPs Unit Officer Course. Majors attended the Military Assistance Officer Command and Staff Course. Language training and civilian graduate schooling in anthropology, economics, foreign affairs, government, international relations, political science, psychology, public administration, or sociology was an integral part of the program.[188]

In April 1970, the Army Research Office contracted a study entitled "Operational and Training Requirements of the Military Assistance Officer." The study, published in May 1971, was based on a review of literature and a series of questionnaires and interviews of the staff and faculty of the Military Assistance Officer Command and Staff Course and the MAOP students from the first three courses.[189] At the time of this study, the MAOP had 552 positions that were classified as direct advisory; other advisory; attaché; area specialist/monitor; CMO (G5/S5); and other functional PSYOP, CA, HQ DA staff, and service school instructors.[190] The skills direct advisors identified as necessary, in the order identified, were language, military professionalism, intercultural communication, resource management, and knowledge of host country forces.[191] Suggestions on intercultural communication included "understanding and respecting the traditional ways and procedures of the host nation military—particularly the matter of pacing, which can be infuriating to the nonreflective member of an Army founded on the maxim of 'do it now'"; "seeing the problems and prospects through the eyes of the counterpart, a vision that requires a sympathetic understanding of his cultural premises and experience"; "abandoning the idea that one's own national/cultural ways of doing things always work best, everywhere and in all circumstances—perhaps a natural temptation of advisors who come to host countries bearing gifts and good advice"; and "the need to tailor one's advice to the culture and conditions of the host country were expressed most succinctly [as] . . . 'to fit in and listen' rather than to [provide] . . . advice out of a 'brown book' [field manual]."[192] Part II of the study provided "Guidelines for Training" the required MAOP skill sets. The study provided many insights into the challenges of developing a special career program that focused on advising foreign armies in a nation-building environment.

Despite the expectation of the CSA in 1968 that there would be a need for 6,000 MAOP officers, by 1970 only 552 MAOP positions existed. As with other special advisory programs during this period, the US Army found it difficult to attract the quantity of qualified personnel needed. In March 1972, the CSA approved combining the MAOP and the FASP into the Foreign Area Officer (FAO) Management System. As of 30 June 1972, only 433 officers participated in the MAOP and 563 in the FASP. By the end of 1972, roughly 900 positions were identified for the new combined program.[193] With this merger, the attempt to develop advisors with an operational focus to meet the civil-military challenges of stability operations ended.

A Senior Officer Debriefing Report, 1972

Major General John H. Cushman submitted his Senior Officer Debriefing Report in January 1972 after serving 8 months as the commanding general of the Delta Regional Assistance Command. Reflecting back on his service in South Vietnam, 22 months advisory duty in the Delta and two previous tours, he "set forth in a reflective vein certain major views held by me at tour's end."[194] Those views most relevant for advisors included what he called "the need for insight," "the advisor," "through Vietnamese eyes," and "the Vietnamese must do it." Cushman began, "All too often insight is gained too late, and through adverse experience." He expressed regret that individual and collective insight had not been better in Vietnam and offered the following:

> Insight—or the ability to see the situation as it really is— is the most valuable asset an advisor can have. Intellect alone does not guarantee insight. Soldierly virtues such as integrity, courage, loyalty, and steadfastness are valuable indeed, but they are often not accompanied by insight. Insight comes from a willing openness to a variety of stimuli, from intellectual curiosity, from observation and reflection, from continuous evaluations and testing, from conversations and discussions, from review of assumptions, from listening to the views of outsiders, and from the indispensable ingredient of humility. Self-doubt is essential equipment for a responsible officer in this environment; the man who believes he has the situation entirely figured out is a danger to himself and to his mission.

He added that insight became "even more a requirement among the intangibles, nuances, and obscurities of a situation like Vietnam." Acknowledging that military officers were men of decision, he explained that "the reflective, testing, and tentative manner in which insight is sought does not mean indecisiveness. It simply raises the likelihood that the decided course of action will be successful, because it is in harmony with the real situation that exists."[195] Insight became the basis of situational understanding that formed the foundation of relevant, feasible advice.

Given the requirement for insight, careful selection of advisors— particularly senior advisors—was necessary. An effective commander might not possess the qualities to be an effective advisor and vice versa. "A marked empathy with others, an ability to accommodate, a certain unmilitary philosophical or reflective bent, a kind of waywardness or independence, and the like—these are often found in outstanding advisors, but may be frowned on in a troop chain of command situation. While it is entirely possible to find the man who excels both as commander and advisor, these men are too rare, and we need to look for good officers who may not be all-purpose officers." Even when the demand for MACV advisors declined, Cushman saw that the need for good advisors would not decline, and might increase in the near future. Consequently, he recommended that a selection board process identify potential senior advisors and that the Military Assistance Institute at Fort Bragg periodically host orientation seminars.[196]

Cushman believed understanding how his counterpart saw things was critical for an advisor to develop rapport and to make good suggestions. He wrote:

> Of course, the advisor must try to see the situation as it looks through Vietnamese eyes; this is part of the insight he strives for—not simply understanding the way Vietnamese in general look at matters, but also how *his* Vietnamese, his counterpart, does. What are the biases, constraints, pressures, and so on, that make up his real world? In all of this, the American has to understand that he is not Vietnamese. He is only temporarily in the country, and he will be exceptional indeed if in his tour he understands a small fraction of how Vietnamese look at their situation and themselves. But everything he suggests should be tested against the question 'how does this fit into the Vietnamese way?' Furthermore, it is very important to understand 'the way things move' and to take advantage of natural movement.[197]

Advisors were reminded that the "natural inclinations" of the Vietnamese—perhaps just as those of the Americans—sometimes worked against their goals. Solutions to these dilemmas should strive to be as "natural" or "least unnatural" to the locals.[198]

Even though Cushman acknowledged that only the South Vietnamese could solve their problems, he believed advisors still played an important role:

> Probably the hardest thing for an American (even for advisors) in Vietnam to grasp completely is that, if our Vietnamese friends cannot bring this thing off, it is not going to get done. We cannot, and should not, do it for them. . . . It means not simply that 'the Vietnamese must do it.' It also means that we must still try to 'show them how.' The job of the advisor becomes more complex, in that he has to figure out what he has to offer at this stage of the war. He can offer a great deal—analysis, an outsider's critique, plus friendly encouragement—all

aimed at ultimate withdrawal of even this support and the Vietnamese doing almost everything on their own.[199]

Acknowledging that the advisor worked basically alone with his counterpart, Cushman found that the "daily mingling of the counterpart with his view and the advisor with his—is what makes advisorship so interesting, and, when it produces a durable and good result, so rewarding and worthwhile."[200]

Cushman's debriefing report contained many valuable observations gained from his Vietnam advisory experience. Other advisors and senior officers also submitted similar reports. In time, the US Army moved out of the advisory trade and focused on defeating Soviet forces in Europe. In that shift back to large-scale, conventional warfare, advisory work dropped off the US Army mission tasks.

A Study of Strategic Lessons Learned in Vietnam, 1980

In an effort to capture the lessons of the Vietnam experience, the US Army commissioned BDM Corporation to conduct an analytical study of the strategic lessons learned in Vietnam. Released in April 1980 and declassified in March 1981, the report of over 3,600 pages consisted of an executive summary and of eight volumes, one volume organized into two books. Although references to the advisory effort can be found in multiple volumes of this massive study, the advisory effort was specifically addressed in book two, *Functional Analyses,* of volume six, *The Conduct of the War.*

The study indicated that during the initial US advisory effort before 1965, the RVNAF developed a limited capability in conventional warfare. However, it was neither trained nor motivated to target the VC infrastructure, its primary threat through 1964, or to conduct counterinsurgency operations. To fight the VC, RVNAF required increasing US combat support assets. At the same time, police forces, just like RVNAF, were not trained or equipped to operate against guerrilla forces. Military advisors were selected on the basis of MOS, rank, and vulnerably to an overseas tour, not on the basis of language skills or the ability to work effectively with Vietnamese counterparts. Because of a lack of civilians with the proper skills that were willing to serve in a combat zone and the availability of military personnel that could be tasked, the military filled many advisory positions more suited to civilians. From 1965 to 1970, MACV combat unit advisors performed principally liaison duties and the quality of advice varied. After 1967 CORDS advisors contributed significantly to pacification and then Vietnamization. During the entire period, the study concluded that there was a general lack of careful selection of personnel to weed out those professionally or personally ill-suited for advisory duty, a lack of adequate language training, and a lack of thorough preparation before arrival. The short tour made it difficult for advisors to acquire the "wide variety of combat-associated experiences needed to know and understand their counterparts, and to gain the cooperation needed to do the job." Reporting accurately remained a problem because of the complex situation and the need to maintain the trust and confidence of the counterpart. An unfavorable report embarrassed the counterpart, threatening rapport, and failed to meet the US Army expectation of progress.[201]

The study identified three successes and five failures for the advisory effort. The first success, the creation of RVNAF as a regular army, was in itself not an easy or simple task. Second, given that counterinsurgency was not its strong suit, the US effort achieved more success against the North Vietnamese Army (NVA) than it did against the VC guerrillas. A third success was the advisor liaison role in coordinating and deconflicting US and RVNAF combat operations that made it easier to work together and for RVNAF units to receive combat support, logistical, and medical support from American units.[202] Three of the five failures did not specifically address advisors. First, preoccupied with the Korean experience, the military problem was misperceived with a resulting focus on defense against an external threat rather than internal security. A second failure, related to the first, was that the military effort was misdirected into a conventional military to repel an invasion instead of a force capable of counterinsurgency operations. Not learning from others—from the French experience in Indochina and from the South Vietnamese themselves—was the third failure.[203]

"Inadequate care in the selection and training of advisors" constituted a fourth failure. The study described the advisor task as "immense"— working and living in an alien culture; establishing rapport with combat-hardened counterparts; advising "on fighting an enemy they did not know or understand, in a terrain they did not know, with troops they did not know"; dealing with civil and military matters; and developing an understanding and maintaining an objectivity need for accurate reporting both to their counterpart and their American superiors. To accomplish this difficult task, young officers, most with limited military service, no combat experience, and little-to-no language skills, were expected to professionally communicate with older, more-experienced counterparts. Poorly prepared advisors could not get respect from counterparts and could not contribute much. "Inadequate language training . . . handicapped them as it made it difficult to learn rapidly even after they were in country, and tended to isolate the advisor from his environment, cutting him off from valuable sources of information." Because the mere presence of a MACV advisor could undermine the credibility of the counterpart, the study states, "the US should have been intent upon providing advisors only where necessary and then making sure that each advisor was so competent that he enhanced the prestige of the United States and contributed substantially to the war effort."[204]

A fifth lack of success was "failure to make maximum use of our advisors." Longer tours could have made up some for inadequate training and deficiencies in language. A short tour provided little incentive for an advisor to invest extra time and effort into language and cultural studies beyond the short training courses. It ensured that advisors were useful for only a few months; that "experience was not cumulative because each year new advisors were being thrown into the conflict and having to learn all over again the lessons which their predecessor had already learned"; that counterparts were discouraged from developing close relations with an advisor by preventing "good, deep working relationships from developing because, given the cultural differences, 1 year was not enough time even where intentions on both sides were the best"; and that the "establishment and maintenance of a reliable intelligence network [became] difficult because the high need for trust in intelligence work did not have time to develop." In short, "the system of short tours destroyed continuity in the US advisor effort and ensured that it was dominated by amateurs."[205]

Two other lessons emphasized in volume 2, *South Vietnam*, affected the advisory effort. First, the political role of RVNAF was critical. Where it formed the support base for a government, it was vulnerable to politicization. Then military leaders were chosen and promoted based on political loyalty, not on military professionalism. Working in this environment was extremely frustrating for advisors and not very productive without support from above. A second lesson was that there was a "tendency when advising or assisting an emerging nation's Armed Forces to organize, equip, and train them in one's own image." Instead of trying to improve the combat effectiveness of the local military using its institutions, systems, and procedures, trying to change them into a version of oneself tended to create numerous problems and proved a difficult long-term undertaking.[206]

The BDM study's final conclusion and recommendation on the advisory effort was that:

> Any future advisory effort should rely on a cadre of highly trained specialists rather than a massive effort by amateurs. The use of specialists familiar with the history, culture and government of the country in which they are to

serve, fluent in the language which they will have to use, and well trained in advisory techniques would improve the likelihood that the failure of Vietnam could be avoided. Specialists with an understanding of the country to which they are assigned will stand a far better chance of correctly assessing the situation and of prescribing solutions which will address the real problems. Furthermore, such advisors would be more likely to earn the respect of their counterparts and, thereby, to establish a relationship of mutual respect. US prestige and influence can only be enhanced by the employment of fully competent advisors even if their numbers are necessarily limited.

The US military services have demonstrated their professional excellence in training foreign personnel and units in technical skills; they have not performed well in advising in politico-military matters because of their lack of background, training, education, and competence.[207]

As with other studies on advisory efforts, the recommendations were not acted on and few lessons were learned. The US Army found the recommendations no longer relevant to its priorities post-Vietnam. At best, they were lessons for what happened in a past but would not be repeated in the future.

US Army Official Publications, 1975–88

In the 1970s, the Department of the Army published 22 monographs as part of its series of *Vietnam Studies*. These provided an initial look into different aspects of the war by authors directly involved in their topics. Brigadier General James L. Collins, Jr. wrote *The Development and Training of the South Vietnamese Army, 1950-1972*. At the end of his monograph, Collins concluded that:

Although many leadership courses were established and continual emphasis placed on the development of leadership, a serious obstacle was United States own overriding emphasis on establishing "rapport" with its counterparts even at the expense of not accomplishing its mission. Too often advisors did not take firm stands with their counterparts on key issues nor recommend the relief of unsatisfactory commanders for fear that such recommendations would reflect badly on their own abilities. If more advisors had insisted on the relief of ineffective commanders, command positions would have opened up, affording incentive and opportunity for the more junior officers to exercise their leadership ability. The rapport approach is dangerous because it lends itself to the acceptance of substandard performance by the advisor. In any future situation where advisors are deployed under hostile conditions, the emphasis should be on getting the job done, not on merely getting along with the individual being advised.[208]

What Collins failed to address was that this approach worked in Korea, but the situation in Vietnam was different. MACV refused to apply leverage from the top and GVN and RVNAF were unable or unwilling to respond as President Syngman Rhee had in Korea. The bottom line remained that MACV advisors experienced frustration with counterparts and attempts to have a RVNAF commander at any level removed from command proved unsuccessful. Without leverage and support from the top—through both MACV and GVN/RVNAF channels—unit and territorial advisors had no tool other than rapport. And the fact remained, as shown in Korea, that leverage from the top still required rapport—personal trust and confidence—at the lowest levels to make the advisory system work.

In the 1980s, the US Army Center for Military History published two volumes in a three-part history of the US advisory effort. While the first volume covered the period prior to 1960, the second published volume addressed 1965–73. During this later period it stated, "MACV commanders did not expect their field and staff advisors to play a major role in the improvement of the South Vietnamese military forces. As liaison teams . . . they kept . . . American commanders abreast of what their allies were doing and where and when they were doing it." Given the magnitude and length of the advisory effort, "American leaders . . . still had a general belief that the advisory teams would have a direct, personal effect on the thousands of Vietnamese commanders being advised. Perhaps it appeared almost self-evident that the US field and staff advisors ought to have some sort of long-term impact on Vietnamese military leadership, if only through sheer weight of numbers. And perhaps it also seemed self-evident that that impact should be positive, given the high quality of American military personnel, almost all officers and noncommissioned officers, assigned as advisors."[209] Unfortunately, what appeared self-evident to some proved false. No MACV commander seriously considered the means that CORDS Ambassador Komer adopted—forcing the relief of incompetent Vietnamese province chiefs. Without support from above, unit advisors had little impact on their RVNAF commanders and found that "getting someone relieved or replaced was like getting a politician out of office."[210]

The advisory effort rested on weak foundations. Preparation was "minimal"—few advisors developed "useful" Vietnamese-language skills and the brief MATA course provided no more than "an introduction to the problems of working in a foreign culture." Recently promoted as lieutenants, captains, or majors, many advisors lacked experience at their rank in US units, much less for Vietnam service. Most faced "cultural shock from being dropped into a completely alien environment before they could begin to become effective." These challenges were compounded by their almost total dependence on Vietnamese interpreters to communicate with their counterparts. In addition, time in an advisory assignment "rarely lasted more than 6 months with any one unit." Only territorial advisory team members—province, district, MAT—received additional training in South Vietnam and most left their jobs just as "they were beginning to know their way around." The quality of field advisors received "only peripheral attention" by MACV and DA. In any case, advisory teams—made up of officers and senior noncommissioned officers—were "rich" in terms of military education and rank. However, the introduction of US combat forces in 1965 made the advisory effort a lower priority.[211] Advisors were selected based on availability for short-tour overseas duty "over and above the needs of regular US units in South Vietnam." The implication that followed was that "given the limited power of the field advisor, and their brief training and short stints in any one position, the value of possessing a special aptitude for the job was probably unimportant."[212]

The sobering conclusion drawn from this history of the advisory effort was that "the final judgment must be that it was beyond the capacity of one power to reform and reshape the society of another." Whether one agrees or not, among the insights offered from this effort were:[213]

- Advisory duty was "much more complicated than it appeared." To expect better results from advisors, better preparation and longer assignments were required.

- Better guidance providing specific goals and the capability of accomplishing them was required for advisors at all levels. "Superiority in material could not compensate for the lack of a unified command, nor could sophisticated plans and programs make up for the absence of more cohesive military and political objectives."

- At the top, the "cult of optimism" in Saigon and Washington was "self-defeating and . . . only encouraged the continuation of policies and practices that had little hope of success."

Summary

For most of the time in Vietnam, the majority of MAAG-V/MACV field advisors—combat unit and pacification—were selected for advisory duty by MOS and rank. Unlike in Korea, almost all attended short periods of instruction designed to familiarize them with advisory duties in Vietnam and to introduce them to the Vietnamese language and culture. Although advisory teams were authorized down to battalion and district level, just as in Korea, teams remained small, often understrength, and frequently filled by personnel who did not meet the rank or MOS requirements for the position. Advisory duty took second place to US military units conducting combat operations. In addition, combat unit advisors served a 12-month tour, but they were assigned to RVNAF units often for no more than 6 months. Facing the same sorts of advisor-counterpart challenges as in Korea, few understood the language, the culture, RVNAF customs, and the local situation. Under these conditions, developing and communicating suitable, acceptable, and feasible advice proved difficult.

In South Vietnam, the United States faced its largest, longest, and most costly advising effort. It addressed a difficult and complex military situation demanding nuanced responses in military and civil-military matters. It consumed the best efforts of the US military for a generation. However, when all was said and done, the combat capability of RVNAF proved inadequate to withstand the North Vietnamese offensive in 1975. For the US military, "no more Vietnams" meant, among other things, no more advisory efforts on the scale or of the duration of that conflict. Consequently, hard-earned lessons and in-depth analyses disappeared from mainstream US military concerns.

Notes

1. Headquarters, United States Military Assistance Command, Vietnam, "Command History 1967," vol. 1 (11 September 1968), 219. Hereafter HQ, USMACV.

2. Ibid.

3. James L. Collins, Jr., *The Development and Training of the South Vietnamese Army, 1950–1972* (Washington, DC: Department of the Army, 1975), 1–7. This is 1 of 22 Vietnam Studies published by the Department of the Army in the 1970s.

4. Ronald H. Spector, *Advice and Support: The Early Years, 1941–1960* (Washington, DC: US Army Center of Military History, 1983), 295–302. This is the first of a three-part US Army Center of Military History official history of the advisory effort in Vietnam.

5. Ronald H. Spector, "The First Vietnamization: U.S. Advisors in Vietnam, 1956–1960," in *The American Military and the Far East: Proceedings of the Ninth Military History Symposium, United States Air Force Academy, 1–3 October 1980*, edited by Joe C. Dixon (Washington, DC: US Government Printing Office, 1980), 111–113.

6. Collins, 17–26, 31. For an overview of Special Forces in Vietnam, see Francis J. Kelly, *U.S. Army Special Forces, 1961–1971* (Washington, DC: Department of the Army, 1973).

7. Ibid., 32, 46.

8. HQ, USMACV, "Command History 1965," 20 April 1966, 82.

9. Jeffrey J. Clarke, *Advice and Support: The Final Years, 1965–1973* (Washington, DC: US Army Center of Military History, 1988), 145. This is the third of a three-part US Army Center of Military History official history of the advisory effort in Vietnam. Part two, covering 1961–64, has not been published.

10. Ibid., 87–96.

11. William C. Westmoreland, *A Soldier Reports* (New York, NY: Dell, 1980), 331. For a Vietnamese perspective, see Ngo Quan Truong, *RVNAF and US Operational Cooperation and Coordination* (Washington, DC: US Army Center of Military History, 1980).

12. HQ, USMACV, "Command History 1967," vol. 1, 237.

13. Clarke, 184–187.

14. HQ, USMACV, "Command History 1966," 30 June 1967, 452.

15. HQ, USMACV, "Command History 1967," vol. 2, 590–591.

16. Clarke, 391–392.

17. HQ, USMACV, "Command History 1968," vol. 1, 30 April 1969, 233–234.

18. Bruce Palmer, Jr., *The 25 Year War: America's Military Role in Vietnam* (Lexington, KY: University of Kentucky Press, 1984), 178.

19. HQ, USMACV, "Command History 1969," vol. 1, 30 April 1970, IV-26.

20. HQ, USMACV, "Command History 1970," vol. 2, 19 April 1971, VII-67.

21. HQ, USMACV, "Command History 1971," vol. 1, 25 April 1972 VIII-73.

22. Ibid., VII-83.

23. Clarke, 452.

24. James H. Willbanks, *The Battle of An Loc* (Bloomington, IN: Indiana University Press, 2005), 161–164.

25. Clarke, 489.

26. Ibid., 494–496; a reduced American military presence remained in the region: a joint US Support Activities Group in Thailand to coordinate air support if the NVA attacked; a US Defense Attaché Office to oversee a restricted security assistance program; and a small US delegation to the Four-Party Joint Military Commission.

27. Initially designated sector and subsector, the terms province and district were used more frequently. To reduce confusion, province and district are used in this study.

28. Clarke, 372; for "seven division equivalents," see Palmer, 178; for 1968 advisor strength (11,596), see HQ, USMACV, "Command History 1968," vol. 1, 23; for 8.65 division equivalents, ratio of 1970 division equivalents (x) / 1970 advisor strength (14,332) = 1968 division equivalents (7) / 1968 advisor strength (11,596).

29. Spector, *Advice*, 291.

30. Martin J. Dockery, *Lost in Translation: VIETNAM, A Combat Advisor's Story* (New York, NY: Ballantine, 2003), 36–40. This excellent firsthand account of an advisor to the RVNAF 2d Battalion, 33d Regiment, 21st Infantry Division from 1962 to 1963 is available in paperback.

31. Robert H. Whitlow, *U.S. Marines in Vietnam: The Advisory & Combat Assistance Era, 1954–1964* (Washington, DC: US Marine Corps History and Museums Division, 1977), 47, 100.

32. Gerald C. Hickey, *The American Military Advisor and His Foreign Counterpart: The Case of Vietnam* (Santa Monica, CA: The RAND Corporation, March 1965), 57.

33. David Ewing Ott, *Field Artillery, 1954–1973* (Washington, DC: Department of the Army, 1975), 24.

34. Clarke, 57.

35. HQ, USMACV, "Command History 1969," vol. 1, IV-26.

36. Peter M. Dawkins, "The United States Army and the 'Other' War in Vietnam: A Study of the Complexity of Implementing Organizational Change" (Ph.D. diss., Princeton University, 1979), 389.

37. Hickey, 59.

38. HQ, USMACV, "Command History 1967," vol. 2, 592–593.

39. Dawkins, 89 and footnote on 34.

40. Cao Van Vien, et. al., *The U.S. Adviser* (Washington, DC: US Army Center of Military History, 1980), 16.

41. HQ, USMACV, "Command History 1965," 76–77.

42. Dawkins, 29.

43. Vien, 16.

44. HQ, USMACV, "Command History 1968," vol. 1, 234.

45. HQ, USMACV, "Command History 1970," vol. 2, VII-67.

46. Spector, *Advice*, 346.

47. US Army Section, Military Advisory and Assistance Group, Vietnam, *Book of Instructions for U.S. Military Advisors to 42 Tactical Zone III Corps South Vietnam* (1962), IV-B-1.

48. Collins, 36.

49. Bryce F. Denno, "Advisor and Counterpart," *Army* (July 1965), 25–26.

50. Clarke, 60–61.

51. HQ, USMACV, "Command History 1966," 452.

52. Ronald D. Ray, interview by Shelby Sears 21 June 2004 (Virginia Military Institute John A. Adams '71 Center for Military History and Strategic Analysis Cold War Oral History Project) [transcript online]; available at http://www.vmi.edu/archives/Adams_Center/RayR/RayR_intro.asp, 12.

53. Clarke, 321.

54. Ibid., 368.

55. Ibid., 67.

56. Hickey, 59–60.

57. Mark A. Meoni, "The Advisor: From Vietnam to El Salvador" (Master of Military Art and Science Thesis, US Army Command and General Staff College, 1992), 28.

58. HQ, USMACV, "Command History 1968," vol. 1, 234.

59. HQ, UMACV, *rf / pf Advisor Handbook*, January 1971, 17–18.

60. BDM Corporation, *A Study of Strategic Lessons Learned in Vietnam*, vol. 6, *Conduct of the War*, book 2, *Functional Analyses*, April 1980, 12-10.

61. Spector, *Advice*, 293–294.

62. Clarke, 61.

63. HQ, USMACV, "Command History 1966," 471.

64. Clarke, 187.

65. HQ, USMACV, "Command History 1966," 473.

66. HQ, USMACV, "Command History 1967," vol. 2, 592–593.

67. Dawkins, 90.

68. HQ, USMACV, "Command History 1970," vol. 2, VII-79–VII-80.

69. Ibid., VII-67.

70. BDM, vol. 7, *The Soldier*, 2-7–2-8.

71. Ibid., 2-8–2-9.

72. US Army Special Warfare School, "Program of Instruction for Military Assistance Training Advisory Course (MATA)" (Fort Bragg, NC: US Army Special Warfare School, April 1962). Hereafter, USAWS, "POI (MATA)."

73. Clarke, 62.

74. USAWS, "POI (MATA)."

75. William Moriarty, interview by Shelby Sears, 24 June 2004 (Virginia Military Institute John A. Adams '71 Center for Military History and Strategic Analysis Cold War Oral History Project) [transcript online]; available at http://

www.vmi.edu/archives/Adams_Center/MoriartyB/MoriartyB_intro.asp, 4.

76. Anthony C. Zinni, interview by Shelby Sears, 29 June 2004 (Virginia Military Institute John A. Adams '71 Center for Military History and Strategic Analysis Cold War Oral History Project) [transcript online]; available at http://www.vmi.edu/archives/archivecoldwar/Details.asp?ID=9&rform=search, 17.

77. Department of the Army, Field Manual 31–73, *Advisor Handbook for Stability Operations* (Washington, DC: Department of the Army, October 1967), 51–54.

78. Zinni, interview, 6.

79. Ibid., 2.

80. Ray, interview, 2.

81. David M. Dacus, "So Now You're an Advisor," *Infantry* (May–June 1971), 32–35.

82. USAWS, "Program of Instruction for Project 404" (Fort Bragg, NC: US Army Special Warfare School, March 1969).

83. Clarke, 62.

84. John G. Miller, *The Co-Vans: U.S. Marine Advisors in Vietnam* (Annapolis, MD: Naval Institute Press, 2000), 38–39. Co-van was the Vietnamese expression for advisor.

85. William E. Colby, "U.S. Assistance Programs in Vietnam: Hearings Before a House of Representatives Subcommittee of the Committee on Government Operations 19 July 1971" [document online]; available at http://homepage.ntlworld.com/jksonc/docs/phoenix-hcgo-19710719.html, 227.

86. Dawkins, 218.

87. Collins, 51–53; Robert Barron, interview by Stephen Maxner, 21 April 2001 (Texas Tech University Vietnam Archive Oral History Project) [transcript online]; available at http://star.vietnam.ttu.edu/starweb/vva/servlet.starweb?path=vva/oralhistory.web&id=newweboh&pass=&search1=WEBOK%3DYES&format=format.

88. Department of State, "Vietnam Training Center District Operations Course," 1 October 1969.

89. HQ, USMACV, *Handbook for Military Support of Pacification*, February 1968.

90. Meoni, 44.

91. HQ, USMACV, *rf / pf*, 7–9.

92. Spector, "First Vietnamization," 114.

93. Spector, "First Vietnamization," 115.

94. Clarke, 58.

95. Stuart A. Herrington, *Silence Was a Weapon: The Vietnam War in the Villages* (Novato, CA: Presidio Press, 1982), 191. This firsthand account of a district intelligence advisor for Advisory Team 43, Duc Hue District, from 1971 to 1972, is available in paperback as *Stalking the Vietcong: Inside Operation Phoenix: A Personal Account*.

96. BDM, vol. 6, *Conduct of the War*, book 2, *Functional Analyses*, 12–13.

97. Herrington, 191.

98. Ibid., 192.

99. Hickey, xiii.

100. Douglas Kinnard, *The War Managers* (Hanover, NH: University Press of New England, 1977), 82.

101. John Cook, *The Advisor* (New York, NY: Bantam, 1973), 42. This is a firsthand account of a district intelligence advisor for the Di An District from 1968 to 1970.

102. Denno, 27–28.

103. William Shelton, interview by Stephen Maxner, 5 June 2000 (Texas Tech University Vietnam Archive Oral History Project) [transcript online]; available at http://star.vietnam.ttu.edu/starweb/vva/servlet. starweb?path=vva/oralhistory. web&id=newweboh&pass=&search1=WEBOK%3DYES&format=format, 28.

104. Herrington, 23.

105. Ott, 25.

106. Dockery, 93.

107. Cook, 68.

108. James F. Ray, "The District Advisor," *Military Review* (May 1965) 7–8.

109. Denno, 28.

110. Dockery, 24.

111. Eddie Jones, interview by Stephen Maxner, 13 January 2003 (Texas Tech University Vietnam Archive Oral History Project) [transcript online]; available at http://star.vietnam.ttu.edu/starweb/vva/servlet.starweb? path=vva/oralhistory.we b&id=newweboh&pass=&search1=WEBOK%3DYES&format=format, 17–18.

112. Denno, 29.

113. Jones, interview, 34.

114. Herrington, 192.

115. Ibid., 193.

116. Clarke, 57–58.

117. Barron, interview.

118. Jackie V. Wright, interview by Dr. Richard Verrone, 12 December 2002 (Texas Tech University Vietnam Archive Oral History Project) [transcript online]; available at http://star.vietnam.ttu.edu/starweb/vva/ servlet.starweb?path=vva/ oralhistory.web&id=newweboh&pass=&search1=WEBOK%3DYES&format=f ormat, 19.

119. Miller, 178.

120. Jones, interview, 22–23.

121. Cook, 101.

122. Hickey, 14.

123. BDM, vol. 6, *Conduct of the War*, book 2, *Functional Analyses*, 12–17.

124. Miller, 182.

125. Hickey, 14–16.

126. BDM, vol. 6, *Conduct of the War*, book 2, *Functional Analyses*, 12–19.

127. Denno, 26.

128. Ibid., 34.

129. Cook, 161.

130. Moriarty, interview, 8.

131. Miller, 177–178.

132. Dockery, 193.

133. Cook, 102.

134. Denno, 29.

135. Spector, *Advice,* 379.

136. Clarke, 69.

137. Irwing C. Hudlin, "Advising the Advisor," *Military Review* (November 1965), 95.

138. Jones, interview, 28, 37.

139. Wright, interview, 19.

140. Cook, 101.

141. David Donovan, *Once A Warrior King: Memories of an Officer in Vietnam* (New York, NY: Ballantine, 1985), 134.

142. BDM, vol. 6, *Conduct of the War,* book 2, *Functional Analyses,* 12–14.

143. Clarke, 63.

144. Miller, vi.

145. Truong, 173.

146. Vien, 74–75.

147. Truong, 166–167.

148. Ibid., 163–164.

149. Vien, 76.

150. Ibid., 71–72.

151. Ibid., 31–32.

152. Truong, 170.

153. Vien, 18.

154. Ibid., 69–70.

155. Ibid., 57.

156. Truong, 164.

157. Vien, 58.

158. Ibid., 151–152

159. Truong, 175.

160. Ibid., 175–176.

161. Vien, 142.

162. Ibid., 152.

163. Ibid., 155.

164. Ibid., 193–194.

165. Ibid., 73.

166. Ibid., 61–62.

167. Ibid., 196.

168. Ibid.197–198.

169. Truong, 174–175.

170. Collins, 130.

171. Hickey, 1.

172. Ibid., 2–3.

173. Ibid., v.

174. Ibid., vi.

175. Ibid., xii.

176. Ibid., xii–xv.

177. Ibid., xvi–xviii.

178. Clarke, 63.

179. Ibid., 64.

180. Ibid.

181. Edwin E. Erickson and Herbert H. Vreeland, 3rd, *Operational and Training Requirements of the Military Assistance Officer* (McLean, VA: Human Sciences Research, Inc., May 1971), 10–11.

182. Ibid., 11–15.

183. John S. Brown, "The Vietnam Advisory Effort," *Army* (March 2006), 95.

184. Erickson, 15–16.

185. DA, "Department of the Army Historical Summary: FY 1971" [document online]; available at http://www.army.mil/cmh/books/DAHSUM/1971/chIV.htm, 36.

186. Department of the Army, AR 614-134, *Military Assistance Officer Program (MAOP)* (Washington, DC: Headquarters, Department of the Army, 20 June 1971), 1–1.

187. DA, AR 614-134, 1-2.

188. Ibid., 1-3–1-4.

189. Erickson, x, 8.

190. Ibid., 18–19.

191. Ibid., 47.

192. Ibid., 57.

193. DA, "Department of the Army Historical Summary: FY 1972" [document online]; available at http://www.army.mil/cmh/books/DAHSUM/1972/ch02.htm, 23.

194. Headquarters, Delta Regional Assistance Command, Vietnam, "Senior Officer Debriefing Report of Major General John H. Cushman, RCS CSFOR-74" (14 January 1972), 1. Hereafter referred to as Cushman Report.

195. Cushman Report, 2.

196. Ibid., 2–3.

197. Ibid., 3.

198. Ibid., 4.

199. Ibid., 5.

200. Ibid., 4.

201. BDM, vol. 6, *Conduct of the War*, book 2, *Functional Analyses*, 12-26–12-27.

202. Ibid., 12-21–12-22.

203. Ibid., 12-22–12-23.

204. Ibid., 12-24.

205. Ibid.12-24–12-25.

206. BDM, vol. 2, *South Vietnam*, 5–56.

207. BDM, vol. 6, *Conduct of the War*, book 2, *Functional Analyses*, 12–17.

208. Collins, 129–130.

209. Clarke, 508.

210. Ibid., 509.

211. Ibid., 510.

212. Ibid., 511.

213. Ibid., 521.

Chapter 3

"A Particularly Tricky Business"[1]

Operations, Plans and Training Teams (OPATT) in El Salvador (1984–92)

> We discovered a combination of not knowing the lessons
> we should have learned from past experience on one hand
> and having to adapt ourselves to somewhat different and
> new situations on the other. It was a tragedy that there
> was no respectable body of doctrine to be drawn on,
> that we were thrown back onto pragmatism. We had no
> respectable organizational approach to deal with this.[2]
>
> —Ambassador Thomas Pickering

Ambassador Pickering's description of the US diplomatic mission's
ad hoc response to the civil war in El Salvador applied equally to US
counterinsurgency (COIN) doctrine and to the US advisory effort at the
time. The 55-man US Military Group (MILGROUP) in El Salvador oversaw
the expansion and training of the El Salvador Armed Forces (ESAF) from
an untrained and poorly equipped force of 11,000 that routinely abused its
authority to a more capable force of 56,000 that could engage its enemy
without alienating the population.[3] The MILGROUP provided advice on
combat operations and on counterinsurgency programs. For many, this 12-
year effort in El Salvador has been viewed as a model for a sustained,
small-scale effort that succeeded. Although neither side won, a peace
settlement was reached in 1991.

MILGROUP and the El Salvador Civil War

A military coup on 15 October 1979 removed General Carlos Romero
from power and served as a catalyst for civil war. The ESAF was an 11,000-
man force officered by tight-knit graduates of the EL Salvadoran Military
Academy and manned by peasant soldiers.[4] A typical Latin American military of that time, it was poorly trained, inadequately equipped, and spread
throughout the country performing security and garrison duties. The
ESAF had a reputation for brutal suppression of internal threats and for
involvement in "death squads" that assassinated dissidents. El Salvador
was divided into six military zones that coincided with territorial divisions of the country. Each military zone, commanded by a colonel who
was designated a brigade commander, was divided into three departments
commanded by lieutenant colonels who commanded two or three battalions of various sizes. Traditionally, the department commanders operated

autonomously from the brigade commander.[5] Through 1980, the ESAF and divergent political groups formed a series of juntas attempting to deal with the massive social, economic, and political problems facing El Salvador. No group proved prepared to handle the problems of government as the "issues addressed and decisions made were always tactical and short-term in nature—the typical bureaucratic 'in-box drill' of finding a 'quick-fix,' and selling it and getting rid of the immediate problem."[6] By October, in frustration, five guerrilla groups formed the *Farabundo Marti para Liberación Nacional* (FMLN) movement in preparation for military action. The United States, fearing "another Nicaragua" with support from Nicaragua, Cuba, North Vietnam, and the Soviet Union, reinstated economic aid to the Government of El Salvador (GOES) on 17 December.[7]

Anticipating a quick victory, the FMLN launched what it called a "Final Offensive" on 10 January 1981. Despite expectations, the "Final Offensive" failed. Unexpectantly, the ESAF "with all its problems and shortcomings" proved better than its opponents. It "forced the FMLN . . . [into] its mountain bases where it spent . . . 6 months analyzing its mistakes, undergoing a process of self-criticism, and improving its organizational unity."[8] In July, the FMLN renewed its attacks, focusing primarily on infrastructure and economic targets. From May to December, the ESAF conducted a series of sweeps, averaging 3 a month and employing 1,500 to 4,500 soldiers. Despite its combat operations, ESAF results were minimal.

On 14 January 1981, immediately after the FMLN attack, the United States restored military aid to El Salvador and sent several advisory teams to assess and address the situation. To improve the capability to take the fight to the insurgents, a US Special Forces MTT from Panama trained the 600-man Atlacatl Immediate Reaction Battalion in El Salvador. In March, for political reasons, the MILGROUP was capped at 55 personnel— officially designated as trainers and forbidden by law from participating in combat operations. With MILGROUP estimates that a cadre of 50 to 60 was the minimum required to train another battalion in country, other options were needed for training ESAF battalions. After rejecting Panama as a training site, the MILGROUP developed a plan to train the Ramon Belloso Immediate Reaction Battalion in the United States at Fort Bragg. Completed in early 1982, the training required a 180-man cadre of US trainers and cost $8 million—enough to train six to eight battalions in El Salvador.[9] When the training of the Atonal Immediate Reaction Battalion in El Salvador was cut short for operational reasons, it became apparent that a regional training facility—eventually in Honduras—was needed to meet ESAF expansion and training requirements.[10]

In the fall of 1981, the US Southern Command (SOUTHCOM) sent a seven-man El Salvador Military Strategy Assistance Team led by Brigadier General Fred E. Woerner to work with the ESAF General Staff to develop their national military strategy. Woerner saw his task as guiding the ESAF leadership in the development of a national military strategy while providing the US government a military assessment of the situation in El Salvador and while conceptualizing a multiyear military assistance program.[11] After 8 weeks' work, a two-part strategy emerged. First, the preparation and training required expanding the ESAF by 10 battalions (8 infantry and 2 quick reaction) to a total of 25 battalions; improving the training base, the command, control, communications and intelligence system, and the combat service support system; modernizing the El Salvadorian air force fixed wing and rotary assets; and increasing navy patrol boats. Second, the ESAF was to conduct "aggressive, small unit, day and night operations" with eight of the new battalions stationed in threatened areas while other ESAF battalions focused on the protection of the upcoming electoral process and the economic infrastructure. The report identified the need for an expensive and long-term American commitment. It noted that the ESAF "has a remarkable capacity for tolerating unprofessional and improper conduct which does not threaten the institution." The report further warned, "Unabated terror from the right and continued tolerance of institutional violence could dangerously erode popular support to the point wherein the Armed Force would be viewed not as the protector of society, but as an army of occupation. Failure to address the problem will subject the legitimacy of the Government of El Salvador and the Armed Force to international questioning."[12]

By the end of 1981, the basis of the MILGROUP security assistance program was in place. Working within the 55-trainer limit, ESAF was expanded over time from 11,000 to 56,000. The guiding principle could be explained as KISSSS, "Keep it simple, sustainable, small, and Salvadoran."[13] It was to be trained by American military personnel and equipped with modernized ground and air assets. Eventually, new units were trained in Honduras and 500 officers for the fivefold increase of ESAF attended officer candidate training in the United States. Training and expanding ESAF, however, was not the principal MILGROUP goal. Under the "rubric of professionalization," that goal was to change the military tradition within the ESAF—a most difficult task. The goals for professionalization—not different from those required elsewhere in the developing world—were an ESAF that (1) subordinated itself to civilian authority, (2) respected human rights, and (3) institutionally changed "so that talent was nurtured, success was rewarded, incompetents were weeded out, and the officer corps in general became operationally effective."[14]

From 1981 to 1984, the ESAF struggled to survive, to expand, and slowly to regain the initiative. In January 1982, 70 percent of the El Salvadorian air force was destroyed in an attack on the Ilopango air base. FMLN forces, operating in strengths up to 600 insurgents, besieged several cities and controlled large areas in El Salvador.[15] ESAF responded with multiple battalion operations—conventional sweeps that did not always differentiate between insurgents and noncombatants. An attempt to execute its National Campaign Plan, a comprehensive pacification program, in two key provinces failed. By mid-1983, ESAF combat effectiveness and morale had improved—the result of new leaders, newly trained units, new training faculties, and better interservice coordination. However, combat operations remained the ESAF focus. It was "still preoccupied with killing the guerrillas and little understood that this aspect of the war was incidental to winning popular support and ultimately the war."[16] American concerns over continued ESAF human rights abuses prompted a visit from Vice President George H.W. Bush whose frank exchange about continued US support depending on a better human rights record prompted a reduction in abuses. By the end of 1984, the better trained 42,000-man ESAF had regained the initiative. With the focus shifting from expansion and training new ESAF units to small-unit counterinsurgency operations and to pacification and civic action, the MILGROUP deployed three-man OPATTs to work with ESAF brigades in 1984. The ESAF expansion had solved its quantity problem; it had enough personnel to stalemate the FMLN. However, it had compounded its leadership problems. In retrospect, "the lesson here is that you have to manage the force expansion very carefully."[17] In addition, officers trained in the United States returned with different ideas than those in the ESAF. Unfortunately, on return they tended to revert to who they were—ESAF officers.

Early in 1985, the FMLN acknowledged the ESAF's improved combat effectiveness by reverting to small-scale guerrilla operations. No longer provided a large, fixed target, the ESAF resisted the need to revert to small-unit operations. In 1986, the ESAF launched its United for Reconstruction civic action plan and Operation PHOENIX to destroy the insurgents near the Guazapa Volcano. Neither was particularly successful by 1987. Even at its peak strength of 56,000, the ESAF "was still not big enough to fight the guerrilla *and* implement an essentially social-economic-psychological operations program at the same time."[18] When an earthquake killed over 1,000 in San Salvador on 12 October, ESAF units were diverted from military duties to conduct humanitarian assistance. Although insurgents targeted MILGROUP personnel, other than four USMC personnel killed at a restaurant when off-duty, the first OPATT member was killed on 31 March 1987 when the 4th Brigade compound at El Paraíso was attacked.

A stalemate between the ESAF and the FMLN ensued from 1987 to 1990. In August 1987, President Jose Napoleon Duarte signed the Central American Peace Plan and then followed up with a broad amnesty program. Off and on attempts at negotiations and an upsurge in human rights abuses characterized this period. These prompted visits and threats from the US Secretary of State in 1988 and Vice President Dan Quayle in 1989. In frustration over the lack of progress in reaching a settlement, on 11 November 1989 the FMLN launched attacks in the cities. Despite almost a decade of work, the surprise attack exposed a failure of the ESAF intelligence in predicting the attack, the ESAF ineffectiveness in responding to the FMLN attacks, and the continued abuse of human rights by ESAF units. In 1990, the United Nations became involved in trying to reach a settlement between GOES and FMLN. That same year, an assistant secretary of defense delivered a "scathing lecture" to the ESAF leadership on the deaths of Jesuit priests and the US Congress cut funds by 40 percent. A helicopter crashed with three US trainers aboard in 1991. One died in the crash; the two survivors were executed. On 17 January 1992, the Chapultepec Peace Accords were signed between GOES and the FMLN. On 1 February, a 9-month cease-fire went into effect and by the summer of 1993, no OPATT personnel remained. On 15 December, a ceremony commemorated the ending of the civil war.

In the end, ESAF had averted defeat, but success against the insurgency had proved elusive. The ESAF had "probably become Central America's most formidable military force, [and] with much pride they argue[d] that if Nicaragua ever started a war, the Salvadorians could finish it. . . . [That] may be accurate . . . but also irrelevant" since it could not defeat the FMLN, its actual threat.[19] Despite the MILGROUP attempts to change the ESAF, its "incongruous approach . . . to organizing and equipping the Salvadoran armed forces in a general conventional manner . . . complicated the task of persuading them to adapt relevant tactics and force structure to the counterinsurgency."[20] It proved impossible to professionalize the ESAF to American expectations.

Combat Unit Trainers

Whether a good thing or a bad thing—and there are strong opinions on both sides of the question— the 55-man limit on the MILGROUP created major challenges. During the early years, the majority of the MILGROUP personnel worked issues at the higher national levels. Temporary American mobile training teams accomplished battalion training. Once the training ended, the MTT left. Not until 1984 were MILGROUP trainers deployed at brigade level in OPATTs.

OPATT Structure

In late 1983, Colonel Joseph Stringham, MILGROUP commander from 1983 to 1984, recommended the establishment of three-man OPATTs at the six brigade headquarters. Each team consisted of a combat arms lieutenant colonel team chief and a combat arms captain training officer, both serving 1-year tours. The third team member was a military intelligence officer serving a 6-month TDY. Each OPATT was made up of US Army personnel except for the 6th Brigade in Usulutan which had USMC personnel.[21] The backgrounds of the six OPATT team chiefs were one Special Forces, three infantry, one military police, and one USMC.[22] At the end of summer in 1984, in the face of ESAF brigade commanders' resistance to lieutenant colonel OPATT chiefs, Colonel James J. Steele, MILGROUP commander, reassigned them to other positions. At the same time, a shortage of Spanish-speaking military intelligence captains prevented their continued participation after the initial 6-month TDY. Thus from late summer 1984 until mid-1985, the OPATT consisted of a captain combat arms training officer. A revised OPATT organization, introduced in mid-1985, had a combat arms major—Special Forces preferred—team chief and two Special Forces warrant officers or noncommissioned officers with training, operations, and intelligence experience. They all served 1-year tours.[23] By the summer of 1991, some of the noncommissioned officer positions were eliminated and in the summer of 1993 OPATTs were abolished.[24]

OPATT Roles

From the beginning, all MILGROUP personnel were designated trainers, not advisors. However, as one OPATT member noted, "the word 'advisor' . . . is more accurate and is a direct translation of the Spanish 'asesor,' which is what we are called by our Salvadoran colleagues."[25] When initially employed in 1984, the OPATT role was to prosecute the war "more aggressively and more humanely." An additional short-term requirement became monitoring ESAF activities during the May elections.[26] After 1985, the reorganized three-man OPATTs focused on ESAF brigade staff operations to improve coordination of operations and intelligence activities with an emphasis on civil defense, civic affairs, and psychological operations. In 1990, monitoring and reporting suspected human-rights violations formally became part of the OPATT mission.[27] By July 1991, being a human rights monitor was a major duty for OPATT.[28] After the signing of the peace plan in December 1991, the OPATT role was described as follows: "Presently, all training activities are restricted to the cuartel [garrisons] and must be coordinated with United Nations Observers. Advisory assistance is now more concerned with civic action, psychological operations and garrison operations and the development of peacetime unit training management systems. Peacetime training activities should increase in the future as the FMLN is demobilized and disarmed. Possible human rights incidents continue to be monitored and reported."[29] In 1992, an OPATT chief wrote in his after action review, "'Observer' would be a far more accurate term than 'advisor' or 'trainer.' These latter two terms require either a willingness of the host nation to accept advice/ help, or lacking that, some sort of power base from which to implement change in spite of local resistance. Neither of these conditions existed for me and so during almost my entire tour, I was strictly an observer."[30] Despite these comments, few OPATT members remembered receiving what could be called a formal mission statement. Most understood their job as improving the combat effectiveness and trying to influence the human-rights performance of their ESAF brigade.

OPATT Selection and Tour Length

OPATT requirements were for 18 trainers—15 US Army and 3 USMC. After 1985, meeting the US Army requirements for 10 Special Forces-qualified warrant officers and noncommissioned officers with regional expertise; training, operations, and intelligence experience; and Spanish language skills proved relatively simple. Most were filled by fully qualified volunteers from the 3d Battalion, 7th Special Forces Group. Finding qualified team chiefs proved more difficult. The MILGROUP requirements for an OPATT chief were combat arms—Special Forces, Infantry, or Armor after Special Forces branch was created in 1987—major or promotable captain, advance course graduate, successful line-company command, battalion or higher staff experience, and professional proficiency (R3/S3) language skill. Special Forces-qualified officers were preferred, but not available in sufficient numbers. In 1990, Special Forces branch had only 17 officers qualified—most were in El Salvador, had just returned from El Salvador, or were assigned elsewhere. The officer manpower pool proved inadequate to meet the need, even when the language requirements were reduced to a limited working proficiency (R2/S2). Even though MILGROUP preferred Special Forces-qualified officers, not until after 1991 were OPATT chiefs coded for just Special Forces officers.[31]

Although the MILGROUP commanders were specially chosen and well-qualified for their assignments and further advancement, the same was not true for all MILGROUP members. Meeting the limited 55-man requirement with fully qualified personnel proved difficult. El Salvador may have been a SOUTHCOM priority, but it was not the priority of the US Army. A 1988 study noted that although superb noncommissioned officers with strong Special Forces experience and extensive regional background and "a few younger officers—energetic and passionately committed to 'their' war—serving on the Operations, Plans, and Training Teams (OPATTs) . . . [were] anything but mediocre," the comments of a former MILGROUP commander that "we had the third team here" may have been too harsh, but it agreed that "the first team went elsewhere." [32] Volunteers were sought, but MILGROUP took what it was given—some against their will and others who had threatened to retire.[33] Even the MILGROUP commander position could prove difficult to fill. An officer indicated that he was offered the job after five others had refused it.[34]

During the initial MILGROUP effort, many MTTs rotated into El Salvador on TDY and others served in Honduras for 89-day or 179-day tours. These limited duty assignments did not stress that, as one SOUTHCOM commander stated, "It's a wartime assignment."[35] Ambassador Pickering "had an enormous problem with . . . longer tours for military personnel. I felt that we were constantly running people through there who had to relearn. The 1-year tour did not become effective for 4 to 6 months, and it was a tragedy that we did this. We didn't have that many people who wanted to come, first, and secondly, we didn't have that many people who could pick up as rapidly on what their predecessors had done, so in a sense we were constantly relearning old lessons."[36] A 1-year tour was better than a 6-month tour, but it proved far from effective. A MILGROUP military intelligence trainer indicated that it took "3 to 6 months for a new advisor to adequately familiarize himself with the enemy situation and the history of the conflict. Consequently, the MI trainer/advisors functioned in a limited capacity 3 to 6 months of a 12-month tour."[37] An OPATT chief stated, "The major problem affecting the OPATT mission was and still is the continued use of a 1 year, unaccompanied rotation of personnel."[38] Short tour lengths—just as in Korean and South Vietnam—reduced the time available for getting a handle on the local situation, for developing rapport with a counterpart, and for developing a long-term approach to the task.

OPATT Preparation and Training

In 1992, a US Army Command and General Staff College student with 11 years Special Forces experience in Latin America and service in El Salvador as an OPATT chief wrote, "Through all of these assignment experiences, I never experienced a formal preparation and training program that specifically addressed how to effectively interact with host nation officers and soldiers as an advisor."[39] Although personnel with a Special Forces background and Spanish-language capability were sought by MILGROUP, no formal training before arriving in El Salvador was provided other than a 2.5-day general, non-El Salvador specific Security Assistance Team Training and Orientation Course (SATTOC). The course did not provide any details on the US military situation in El Salvador or on GOES or ESAF. It offered nothing on advisor duties, no ex-advisors were on the staff, and there were no role-playing exercises or case studies. A student described the training in 1989 as "very close to completely useless." Another noted, "All this time the advisor had expected that, somewhere along the line, he would be briefed on his role as advisor, counterinsurgency, the Salvadoran National Plan, etc. . . . But this doesn't happen."[40] The assumption was not that anyone could be a trainer, but that anyone with regional experience, particularly Special Forces, and with Spanish language capability could.

The initial briefings for the newly arrived OPATT members in El Salvador were not much better. In 1986, an OPATT chief "read two three-ring binders of policy . . . signed that I read them . . . left for Santa Ana . . . I didn't meet . . . [MILGROUP commander] during my first 100 days in El Salvador." Another described briefings that focused on "telling me what not to do . . . nothing about what to do."[41] Another was told, "Do nothing that will jeopardize the US Security Assistance Program here. You will probably not greatly effect [affect] the status of your unit while you are here, so don't try. Look for the small victories that may have a cumulative effect."[42]

Challenges of the Advisory Environment

Even with OPATT trainers with Spanish language skills, some cultural sensitivity, and service in the region, there remained what an OPATT chief called the "fundamental problems in the advisory business."

- Advisors are brought into the country by the national level host nation [HN] for their own reasons. "This does not mean that the local people and lower levels want advisors or advice."

- Host nation forces are often corrupt to one degree or another. They don't want anybody looking at their operations, finding out about certain types of activities, reporting on problems, or fixing money issues.

- Often, the advisor (regardless of rank) could do everybody's job in the HN unit—better than the person holding the job. The advisor has all kinds of ideas to improve things. To the counterpart, this is a threat.

They "will generally be hesitant to invite scrutiny because, however nicely it may come out, there will almost always be some form of criticism. If someone can prevent an inspection of their operation, they will."

- There are basic misunderstandings between what the advisor thinks he should be doing or what he is capable of doing and what the host nation thinks the advisor should be doing or wants him to do. Getting this straight is an important task for the host nation and top-level US authorities to agree on. "Written guidance is needed to make sure this is all perfectly clear to all parties. If left to the HN, they will so design the advisor's job that he is no threat, provides only resources—without specifically controlling them—and does not see anything which reflects badly on the HN unit or personnel."[43]

For each OPATT member, even the well qualified, these basic problems meant "advising remains a particularly tricky business. Every advisor is placed in the tricky position of trying to influence the behavior of others over whom he has no authority, causing them to do things that may be foreign to their nature and habit, while at the same time attempting to interpret, implement, and respond to criticism of the US political decisions over which he had no input or control. Furthermore, all of this occurs against the backdrop of severe social, institutional, and political stress that is inherent in societies in conflict."[44] In other words, "advisory duty will continue to be one of the most vexing, enriching, challenging, and memorable tours available."[45]

Understanding: Culture and Language

The US military had many personnel with experience in Latin America. Although alike in many customs and most sharing a common Spanish past, each Latin American country had developed in its own way. Although many appeared similar from a distance, each country was unique up close. An early MILGROUP commander noted, "You can't look at El Salvador unless you understand the *Matanza*. You can't look at El Salvador unless you understand the impact of the Soccer War. You can't look at El Salvador unless you understand the population pressures, land distribution, the pre-revolutionary situation. . . . You have to do your homework."[46] And as always with advisors at the lowest levels, one OPATT member noted that "cultural immersion is total and the pace is intense."[47]

Language proficiency for OPATT members was well above that of Korean and Vietnam advisors. However, then as now, Army Regulation 12-15, *Joint Security Assistance Training (JSAT)*, stated, "A request for team members with foreign linguistic ability can rarely be honored. Necessary interpreter support will be the responsibility of the foreign country. MTT requests may indicate that language capability is desired but will not state a mandatory requirement."[48] After years of assistance training and advisory work, providing personnel with appropriate language skills was not mandatory. Interpreters, provided by the host nation, were the standard. However, as one OPATT trainer observed there was "little excuse for the US military not to assure the '55' advisors sent annually to El Salvador are linguistically proficient."[49] Even so, the required language skill levels for OPATT chiefs were lowered because of a lack of available qualified personnel and at least one MILGROUP member was sent home for inadequate language skills. Even those who spoke some Spanish indicated the need to continue to learn more. An OPATT chief offered, "Language Training, Language Training, Language Training! What is the minimum acceptable level (1-1, 2-2, 2+-2+)? . . . Non-native speakers should . . . be funded to continue formal language training while in country."[50] In addition, another believed, "Advisors must be language proficient at the native speaker level. They must be experts in LIC (Low Intensity Conflict) doctrine, weapons employment, and small unit tactics. And they must be diplomats in every regard."[51]

Developing Rapport with Counterpart

Without an understanding of the counterpart and his problems as he viewed them, improved language skills and cultural awareness counted for little. OPATT members needed to develop an understanding of the ESAF—its expectations, how it worked, what it was capable of doing, and what it was unwilling to do—and of the problems it faced—political, social, economic, and security—to be able to develop rapport with their counterparts and to provide useful advice based on actual situational understanding. This was not simple. As a military attaché observed:

> We're making a lot of assumptions which are incorrect. We assume that [an officer] has certain basic leadership skills, that he's going to check on his guys, that he's going to go out himself and make sure that things are being done, when that's probably not the case. . . . We assume all these things and we teach at a higher level. The fact is that these officers don't operate that way . . . [and] the training that we give the young soldiers and the young cadets won't get employed because they immediately forget all of the good things they've learned and adapt to the bad habit of their own chain of command.[52]

As in many countries, the officers were not well paid and there was a tradition of corruption. The system reinforced the importance of personal friendships and family ties. There was a "hesitancy to do anything that's confrontational." Contrary to American expectations, senior officers did not "follow through and make sure that his order has been executed according to his intent. The sense of urgency and obligation to do that does not exist. It's not there."[53] Brigade, department, and battalion commanders often operated independently from one another and were protective of their prerogatives.

The ESAF wanted to defeat the FMLN, but it was careful not to jeopardize its position in El Salvador. It supported the expansion from 11,000 to 56,000 personnel. Equipped and trained conventionally, it sought to destroy the FMLN using battalion or larger operations depending on artillery and air support financed by American military aid. "Gringo" training was necessary because of funding, but it did not change the institutional ethos and goals of the ESAF. Before the civil war, the "tanda" or military academy system produced 30 officers each year.[54] They were promoted by year group and were practically guaranteed retirement with the rank of colonel. Group loyalty was stressed over military competence and honesty. Because the ESAF consisted of officers and of illiterate peasant soldiers, there was no basis or acceptance for a noncommissioned officers corps that was seen to threaten the authority of the officers. Without the small-unit leadership provided by noncommissioned officers, ESAF resisted the MILGROUP pressure to conduct small-scale counterinsurgency operations.

Often brigade commanders did not welcome OPATT assistance. According to one lieutenant colonel OPATT chief, "I know that the brigade commanders weren't comfortable with [us] because we were bold enough to ask questions about their plans and operations. And the brigade commanders wanted people who responded . . . instead of asking questions."[55] After 1985, the OPATT priority shifted from "training, which most brigade commanders valued . . . [to] objectives . . .in precisely those subjects that their counterparts cared about the least—which included monitoring and reporting human-rights violations." Rapport—"an ability to communicate and have suggestions well-received"—depended largely on the attitude of the brigade commander. Some were notorious human-rights offenders and others proved just reticent. When brigade commanders remained distant, many OPATT members attempted to work with anyone receptive to their assistance, particularly the operations and intelligence members of the brigade staff. "Clearly, adviser-counterpart relations were not uniform. . . . Personal and professional factors could combine with cultural differences to make for trying circumstances, especially against the backdrop of the drawn-out insurgency."[56]

Some OPATT members came to accept that "assisting armies may involve host-nation political, cultural, economic, or social changes that will take years to complete." Any progress would be slow; nothing happened quickly. Some came to realize that American "attitudes and practices often have very little validity for Third World armies. Thus, an advisor should not measure his progress by US standards but rather by standards that have relevance to the host nation." To understand the situation and to be able to develop practical advice:

> An advisor needs to be aware of the differences between his own military system and that of his host's and to explore all of the reasons that the host nation performs a task in a given manner before recommending a different way to do it. It is also a good idea to consider the reasons why the host nation might not be able to do a given task before it is recommended. Host nation capabilities and limitations must be considered prior to submission of any advice or suggestion. At all costs, the advisor should avoid unfavorable comparisons of host-nation practices with US methods.[57]

When working with a nonresponsive counterpart, it proved difficult for an advisor "to cultivate friendships . . . to influence the institution in positive directions . . . if host nationals are not talking to him." Advisors needed to be aware and sensitive "to the difference in cultural ethics that may exist between the two cultures and ensure that his conduct is above reproach in either of the societies, particularly with . . . cultures that may have distinctly different views and customs about what is financially, morally, or legally acceptable." If "the advisor is a salesman with a worthwhile product that will help immeasurably if the consumer just learns a bit more about it," that product had to meet the counterpart's needs within his resources, his system, and his goals.[58] Another OPATT chief suggested, "Have advisors work on simple things. . . . Philosophy does not count for much, look for practical improvements. . . . Advisors need to regress in their thinking to understand where problems may be. Think very, very basic . . . can everyone see the target? . . . Can the troops read?"[59] Problems could be so basic that they were not even considered.

US Army Pressures: Formal and Informal

Looking back, a former MILGROUP commander noted early ESAF operations of big sweeps and multibattalion operations "looked like a repeat of the American Way of War, namely, bigger, louder and more of the same."[60] Another ex-MILGROUP commander believed that early in the training of units "one of our problems [was] . . . the tendency for us to want to organize their units around how we are organized." This tendency carried over into expectations for leadership, for noncommissioned officers, for combat operations, and for counterinsurgency operations. To win, the ESAF did not have to meet US military standards; it just had to be better than the FMLN. In working through frustrations and the slow, laborious process, MILGROUP members were advised to resist the American three-step—"there is a tendency for Americans to want to do things quickly, to do them efficiently and the third step in that process is to do it yourself." Anything done by the Americans, rather than the ESAF, was "going inevitably to be viewed as a Gringo solution." Without ESAF buy-in, all "Gringo" programs were destined for limited results at best, but more likely failure. And it would be the Americans' fault. A hard-learned recommendation was "to the extent that you can make their institutions work or you can take what they have, whether it's leadership in the form of people or it's organization or even equipment, you ought to do that. It works and it's worth the effort."[61]

Counterpart Observations

Just because the United States provided the resources for the expansion of the ESAF and for fighting the war was no reason for the ESAF leadership to accept American military concepts, particularly if they did not meet ESAF perceived needs. Many commanders resented US interference, particularly the American emphasis on human rights. Others resented "American impatience (some would say arrogance) when confronting a different culture, as well as the 'can-do' inclination to take charge in the face of inefficiency or ineptitude."[62] OPATT members expressed frustration about being unable to accompany ESAF units during operations to assess combat performance and to identify training deficiencies. In fact, an ESAF joke was that "asesor"—the Spanish word for "advisor"— actually meant "one who tries to tell us how to run a war without ever having been there."[63]

Both the ESAF and FMLN acknowledged that OPATTs made a difference in the war. Although the ESAF never fully accepted small-unit operations as a critical component of the security side of counterinsurgency, it developed a national pacification plan and focused to a degree on civic action and on working among the people. However, its execution fell short of its goals because of inadequate ESAF manpower for pacification, security, and combat operations; lack of resources for the pacification programs; and the diverting of assets to humanitarian relief efforts after the earthquake in 1986. One ESAF officer acknowledged, "Our biggest mistake as military men is to assume that there is a quick and rapid solution to this war and that all we need to do is change a few tactics here and there. But this is simply not the case. This type war is new and it's so politically inclined that its entire framework is very different."[64] A former FMLN leader believed that American advisors at the brigades made the ESAF more effective, more professional, and less abusive. To ensure that the ESAF did not change its mind about the peace settlement, the FMLN insisted on the presence of OPATT members with ESAF brigades during demobilization.[65]

Special Studies and Other Observations

Just as with previous advisory efforts, studies and participants analyzed the MILGROUP effort in El Salvador to learn lessons and to understand what happened. What follows are some observations, conclusions, and recommendations from the 1987 comments of Ambassador Thomas Pickering, a 1988 special report completed by four US Army officers, a 1992 OPATT after action report, and a 1995 RAND study.

An Ambassador's Thoughts, 1987

Ambassador Pickering served as the US Ambassador to El Salvador from 1983 to 1985. His observations about US government inadequacies were paralleled by similar US military shortcomings. Facing an insurgency in El Salvador, Pickering discovered that he had:

> . . . neither the doctrine nor the support nor the coordination in the US Government that would really be required to deal effectively with that kind of operation. I don't think we ever developed it. We still are kind of ad hoc in our way of viewing the problem. That is really quite a critical comment. The fact that we were reasonably successful has very little to do with the fact that we had previously developed the answers to those issues, and in effect we were often condemned to reinventing a lot of them. In a way, we were unhampered by doctrinal preconceptions, and that helped in pragmatism and flexibility, but in another sense this made it very difficult to stay the course, to know what other things happened, to do all the things that had to be done all at once.[66]

According to Pickering, the impact of this shortcoming was "first, that in the failure of the United States effectively to study, assess, write histories about, and reach conclusions on these types of wars we are condemned to refight and rediscover them. Secondly, it's very important when we deal with this kind of a conflict that we deal with it on a coordinated basis." Pickering concluded, "The next stage after being able to understand what happened before is to have some distilled wisdom to build on as a result of having taken a look at what has gone before."[67] Today, 30 years later, Pickering's comments still ring true.

The "Four Colonel's" Report, 1988[68]

During the 1987–88 academic year, four US Army lieutenant colonels attending Harvard University's John F. Kennedy School of Government wrote *American Military Policy in Small Wars: The Case of El Salvador*. Published in 1988, the report was based on oral histories held by the US Army Military History Institute, interviews with US and ESAF military personnel, and firsthand observations from a trip to El Salvador in October 1987.[69] The report stated, "The results of American efforts to professionalize the officer corps—and they achieved partial success at best—are instructive."[70] Partial success was acknowledged in the ESAF acceptance of civilian control and in the reduction, but not the elimination, of human rights abuses. However, "attempts to supplant the ethos of the Salvadoran officer corps with a more professional model and to develop noncommissioned officers produced little, despite the expenditure of prodigious resources. Whether either effort ever had much chance of success, given . . . Salvadoran culture and military traditions, is questionable."[71] At the lower advisory levels, rapport based on a close personal relationship by a militarily competent, linguistically capable, and culturally informed advisor still faced another difficult barrier—the host nation military culture and its institutional imperatives, which are always resistant to any change, much less quick change.

The MILGROUP effort to professionalize the ESAF made the least progress in changing the culture or ethos of its officer corps. The chief obstacle to a competent officer corps was the military academy class or "tanda" system that promoted all its members together based on year group. This ensured that "whatever an officer's personal failings—stupidity, cowardice in battle, or moral profligacy—his career [was] secure through the rank of colonel, after which he may depart, with his tanda [year group], into honorable retirement." As a privileged class, ESAF officers lacked "a commitment to technical mastery or a sense of responsibility for the performance of their units . . . [and] concern for the common soldier's welfare." To Americans, the ESAF leadership undervalued training; had a cavalier, unprofessional approach toward combat operations; and did little to improve the soldier's lot, which bordered on neglect. Making limited progress in breaking these habits, the MILGROUP placed its long-term hope on a new generation of officers created by the expansion of the ESAF and trained in the United States. Instead of creating a core of military competence and professionalism, this program created tensions between the "gringo officers," mostly lieutenants, and the more senior ESAF officers. Officers were forced to "choose between being ostracized and . . . adherence to Salvadoran military traditions." Officers who appeared to support the military skills and professional values taught at Fort Benning in the United States or at the Regional Military Training Center in Honduras reverted "to the old way of doing things with disturbing frequency once back in El Salvador." Despite MILGROUP efforts, and as most other armies do, the ESAF viewed "autonomy over its internal affairs as essential to its institutional integrity." The MILGROUP found it much easier to create and train new battalions than to produce competent leadership to command them effectively. [72]

Efforts to create a noncommissioned officers corps in the ESAF proved equally frustrating. Although MILGROUP trainers found "that the NCO concept is alien to the Salvadoran military tradition, as it is throughout most of Latin American," they continued to push the program that made perfect sense to them as Americans. Ignoring the ESAF structure of commissioned officers and short-term peasant conscripts, a noncommissioned officer rank structure was superimposed on the ESAF; one that it did not understand nor accept. Attempting to create a corps of noncommissioned officers in the ESAF underscored "the difficulty of undertaking institutional change that ignores strong cultural biases." ESAF officers did not understand, welcome, or accommodate a noncommissioned officer corps that was considered a threat to the officer corps. Looking back, "the American attempt to create an NCO corps appears naïve and presumptuous." A clear lesson both for US military policy and for advisors is to "concentrate on issues that are not only relevant to a counterinsurgency—as NCOs indisputably are—but also reasonably attainable given the war's specific context. To do otherwise is to risk squandering resources that are already in short supply."[73] Advisory work proved a frustrating and difficult enough challenge with its plethora of misunderstandings and divergent goals without the added handicap of attempting, out of ignorance or willfulness, the unattainable. Working within the host nation military system requires knowledge, thought, understanding, adjustment, and effort; but in the long run, it may prove easier and more effective in the short-term than trying to change the unchangeable.

By focusing on organization, equipment, training, and tactics, the MILGROUP work to change the way the ESAF fought literally transformed the ESAF. Reaching a maximum strength of 56,000, the ESAF was bigger, better trained, better equipped, and seasoned by years of fighting. The expansion created a force ratio that ensured the FMLN would not win. ESAF soldiers were trained, equipped, and capable of living and fighting in the field. However, the ESAF proved incapable of destroying the FMLN insurgents. A major improvement, "perhaps the most spectacular," was in the changes in the quality and quantity of its facilities—command and control, logistical, medical, training, and airfield. However good the progress, there remained disappointments. "Despite the oft-expressed American intent to convert ESAF into a counterinsurgent force . . . it failed to wean the Salvadorans from their conventional mindset. If anything, American actions have reinforced that bias." Organizing and equipping units conventionally with heavier weapons than those suited for counterinsurgency operations reinforced the ESAF preference for conventional, large-unit operations and stiffened their resistance to decentralized, small-unit saturation patrolling more suited to counterinsurgency operations.[74] By replacing damaged fixed and rotary wing aircraft, the MILGROUP produced an El Salvadoran air force capable and willing to support combat operations. The result was that for counterinsurgency operations that emphasized "being *among* the people, the UH-1 . . . made ESAF into an army that spends too much time *above* the people." The report found that "almost instinctively, Americans take a rich man's approach to war."[75]

The authors of the study noted that at the Kennedy School, as well as in America, "the notion that no problem of public policy lies beyond solution has entrenched itself as an article of faith." They continued, "To soldiers, optimism comes less easily, for no historical phenomenon has proven more resistant to simplified prescriptions than the subject of their profession." Unfortunately, the American military was not immune to the belief that problems have solutions and that as Americans they can "make it happen." The study made the following recommendations:

- Make room for the study of small wars in military schools.

- Clarify organization responsibilities for fighting small wars, in Washington and in the field.

- Overhaul the procedures governing security assistance.

- Before undertaking any intervention, establish a vision of what you hope to accomplish and a consensus of political support to sustain that vision.

- Put someone in charge, vesting that official with real authority.

- Send your first team and permit its members the latitude needed to get the job done.

- Foster institutional change only where it will make a difference.

- Avoid introducing inappropriate technology.

- Weight the 'other war' as the tougher part of the proposition.[76]

Although these suggestions appear to be simple and common sense, things always prove more difficult in practice than they do in theory.

An OPATT After Action Report, 1992

On 18 April 1992, the 2d Military Zone/Brigade OPATT Chief, a Special Forces major, submitted his end-of-tour after action report. This 5-page report and its 10-page attachment provided an analysis of problems confronted during a frustrating tour. He offered suggestions and recommendations on how to do the advisory business better; several were referred to earlier in this chapter. Despite working with foreign soldiers previously, he found that the "advisory business and the mission of the MILGROUP are the most difficult jobs I have thus far seen in my career." Addressing the disparity between relationships between advisor and their counterparts, he did not believe that "our doctrinal approach to the advisory business should be based on luck. If the job is worth doing, it is worth doing right and requires planning, organization, and systemic solutions not trusting to the good will of the host nation and hoping they will do what we want." This required common efforts and expectations through both MILGROUP and the ESAF channels. He expressed concern that because of the political settlement in El Salvador "the US Army now believes it knows how to handle insurgencies and establish effective MILGROUPs." He firmly stated, "It does not." He thought that the MILGROUP advisory work in a counterinsurgency environment "presented a far more difficult job and a greater challenge than anything else our army has done in many years. For all the difficulty of conventional operations, they are not even in the same ball-park as far as the need to be innovative, creative, and juggle a host of political, military, social and economic requirements. The fact is that nobody is adequately trained for the work that makes a complex job extremely difficult."[77]

A RAND Report, 1995

In 1995 RAND published *The Effectiveness of U.S. Training Efforts in Internal Defense and Development: The Cases of El Salvador and*

Honduras. This comparative study provides some useful insights for American advisory personnel to consider. First, that US military trainers need "to be more sensitive to the host-nation training needs of Third World militaries . . . commonly heard laments that the United States frequently teaches skills and ideas that are of little use to the Salvadoran or Honduran soldiers. The idea is to teach US doctrine to Third World officers for greater understanding and interoperability between the military establishments. However, much of US doctrine is simply not applicable within the Third World context."[78] Second, the approach of using military training to change a foreign military's internal ethos rests on a weak foundation:

> As the United States has attempted to train the skills applicable to internal defense and development, it has faced three fundamental hurdles. First, can the United States effect basic attitudinal and behavioral change in the *individual* soldier who receives the training? Second, assuming the individual soldier internalizes the lessons on 'professionalism,' can this individual-level metamorphosis be translated into a wide *institutional* transformation? Third, given the multitude of exogenous factors that affect democratic political development and structurally induced political repression, can the military play an instrumental role in effecting change on a *societal* level? Strong political, personal, historical, and financial reasons abound for these militaries to remain politically viable and independent. Consequently, it would appear that no amount of US training could persuade them to do otherwise.[79]

Third, many studies have "concluded that US military training and equipment have little or no effect, negative or otherwise, on the institutional behavior of Latin American militaries."[80] Basic skills and tactics, techniques, and procedures (TTPs) can be improved by American military training as shown in Korea and Vietnam. However, changing the internal military culture of an army proves to be a difficult, time-consuming, and problematic task. While a unique, well-considered, carefully-tailored, and long-term approach might work; normal military training and normal military approaches do not. The advisory environment offers no easy, quick fixes. It demands hard, focused work and an unusual situational understanding.

Summary

Unlike Korea and Vietnam, MILGROUP trainers were selected for their language skills and prior experience in the region, as well as MOS and rank. As a result, no special courses or preparation were considered necessary. Brigade OPATT teams remained small, partly because of the MILGROUP 55-man limit. Just as in Vietnam, trainers were limited in their approach to problems and in their effectiveness by a 12-month tour. Unlike Korea and Vietnam, OPATT trainers appear to have worked more with members within the staff and unit rather than as counterparts to the brigade commander.

Without MILGROUP training, equipment, and advice, the ESAF would have failed. The effort in El Salvador was a long, financially costly affair; but the initial aim of preventing a FMLN victory was met. The creation of an ESAF that was organized and capable of conducting small-unit counterinsurgency operations among the populace proved elusive—both because of the ESAF resistance and the American approach that organized, equipped, and trained ESAF for what it knew best—conventional operations. However, the goal of professionalizing the ESAF—of changing its internal values, customs, and traditions to those resembling a modern professional military—did not happen.

Notes

1. David L. Shelton, "Some Advice for the Prospective Advisor," *Marine Corps Gazette* (October 1991), 55.

2. Ambassador Thomas Pickering, quoted in Max G. Manwaring and Court Prisk, *El Salvador at War: An Oral History of Conflict from the 1979 Insurrection to the Present* (Washington, DC: National Defense University Press, 1988), 244.

3. Max G. Manwaring and Court Prisk, "A Strategic View of Insurgencies: Insights from El Salvador," McNair Paper No. 8 (Washington, DC: Institute for National Strategic Studies, National Defense University Press, May 1990), 12.

4. Andrew J. Bacevich, James D. Hallums, Richard H. White, and Thomas F. Young, *American Military Policy in Small Wars: The Case of El Salvador* (New York, NY: Pergamon-Brassey's, 1988), 5. A chart on this page identifies both the ESAF troop strength and force structure for each year from 1979 to 1987.

5. Colonel Lyman Duryea, US military attaché, quoted in Manwaring and Prisk, *El Salvador*, 298. One military zone had only one department; the other five had three departments each.

6. Manwaring and Prisk, "A Strategic View," 8.

7. For a chronology of events in El Salvador from 1979 to 1991, see Benjamin C. Schwarz, *American Counterinsurgency Doctrine and El Salvador: The Frustrations of Reform and the Illusions of Nation Building* (Santa Monica, CA: RAND Corporation, 1991), 85–92.

8. Tommie Sue Montgomery, "Fighting Guerrillas: The United States and Low-Intensity Conflict in El Salvador," *New Political Science* (Fall/Winter 1990), 32.

9. Colonel John D. Waghelstein, MILGROUP Commander 1982–83, quoted in Manwaring and Prisk, *El Salvador*, 236–237.

10. Mark A. Meoni, "The Advisor: From Vietnam to El Salvador" (Master of Military Art and Science Thesis, US Army Command and General Staff College, 1992), 48–49.

11. Brigadier General Fred E. Woerner, quoted in Manwaring and Prisk, *El Salvador*, 115.

12. Fred E. Woerner, "Report of the El Salvador Military Strategy Assistance Team (DRAFT)" [document online]; available at http://www.gwu.edu/%7ensarchiv/nsa/DOCUMENT/930325.htm.

13. Montgomery, 29.

14. Bacevich, 24–25.

15. Manwaring and Prisk, "A Strategic View," 10.

16. Waghelstein, quoted in Manwaring and Prisk, *El Salvador*, 223.

17. Colonel James J. Steele, MILGROUP Commander, 1984–86, quoted in Manwaring and Prisk, *El Salvador*, 409.

18. Montgomery, 43.

19. Manwaring and Prisk, "A Strategic View," 16.

20. Ibid., 17.

21. Cecil E. Bailey, "OPATT: The U.S. Army SF Advisers in El Salvador," *Special Warfare* (December 2004), 20.

22. Ibid., footnote 18, 28.

23. Ibid., 21.

24. Ibid., 27.

25. Roland B. Waters, "The Marine Officer and a Little War," *Marine Corps Gazette* (December 1986), 54.

26. Bailey, 18.

27. Ibid., 21–22.

28. Meoni, 90.

29. Bailey, 22–23.

30. "After Action Report, 2d MILZONE OPATT Chief, Apr 91–Apr 92," Memorandum for Commander, MILGROUP, 18 April 1992. Hereafter referred to as AAR 2d OPATT.

31. Bailey, 21–22.

32. Bacevich, 16–17.

33. Meoni, 167.

34. Bacevich, 18.

35. General Wallace H. Nutting, SOUTHCOM Commander, 1979–83, quoted in Manwaring and Prisk, *El Salvador*, 242.

36. Pickering, quoted in Manwaring and Prisk, *El Salvador*, 243–244.

37. Victor J. Castrillo, "Contributions, Shortcomings, and Lessons Learned From U.S. MI Training/Advising in El Salvador," *Military Intelligence Professional Bulletin* (October–December 1993), 41.

38. Meoni, 98.

39. Ibid., 4.

40. Ibid., 63, 189.

41. Bailey, 23.

42. Meoni, 195.

43. AAR 2d OPATT.

44. Shelton, 55.

45. Ibid., 57.

46. Waghelstein, quoted in Manwaring and Prisk, *El Salvador*, 242.

47. Shelton, 55.

48. Department of the Army, Army Regulation (AR) 12-15, *Joint Security Assistance Training (JSAT)* (Washington, DC: Headquarters, Department of the Army, 5 June 2002), 171–172.

49. Meoni, 65.

50. AAR 2d OPATT.

51. Jeffrey U. Cole, "Assisting El Salvador," *Proceedings* (November 1989), 69.

52. Colonel Lyman Duryea, quoted in Manwaring and Prisk, *El Salvador*, 303–304.

53. Ibid., 304–305.

54. Waghelstein, quoted in Manwaring and Prisk, *El Salvador*, 278.

55. Bailey, 20.

56. Ibid., 23–24.

57. Shelton, 56.

58. Ibid., 56–57.

59. AAR 2d OPATT.

60. John D. Waghelstein, "Ruminations of a Pachyderm or What I Learned in the Counter-insurgency Business," *Small Wars and Insurgencies* (Winter 1994), 369.

61. Steele, quoted in Manwaring and Prisk, *El Salvador*, 407–498.

62. Bacevich, 23.

63. AAR 2d OPATT.

64. Colonel Carlos Reynaldo Lopez Nuila, quoted in Manwaring and Prisk, *El Salvador*, 281.

65. Bailey, 27.

66. Pickering, quoted in Manwaring and Prisk, *El Salvador*, 245–246.

67. Ibid., 485.

68. Victor M. Rosello, "Lessons from El Salvador," *Parameters* (Winter 1993–94), 106.

69. Bacevich, 2.

70. Ibid., 25.

71. Ibid., vii.

72. Ibid., 26–27.

73. Ibid., 27–28.

74. Ibid., 28–29.

75. Ibid., 29, 31.

76. Ibid., 49.

77. AAR 2d OPATT.

78. Michael Childress, *The Effectiveness of U.S. Training Efforts in Internal Defense and Development: The Cases of El Salvador and Honduras* (Santa Monica, CA: RAND, 1995), xviii.

79. Ibid., 66.

80. Ibid., 12.

Chapter 4

Observations

> In retrospect the role of the US military advisor has not
> undergone an orderly historical evolution during the past
> quarter century. . . . The patterns established in the advi-
> sory effort reflect the influence of circumstances rather
> than of evolution.[1]
>
> —Walter G. Hermes, 1965

From this brief review of the American advisory experiences in Korea, Vietnam, and El Salvador, the above statement from the US Army Center of Military History report, "Survey of the Development of the Role of the U.S. Army Military Advisor," remains valid today—over 40 years later. Each time the US military response to advisory requirements was an ad hoc, secondary endeavor. Each time results were expected. Each time advisors tried their best. Each time the results were mixed. Each time the experience was forgotten—relegated to that lesser important, not-to-be-done-again-anytime-soon pile of military tasks.

What Advisors Did: Organization and Roles

American MAAGs were forced to get the best results out of the minimum number of personnel. At the lowest unit advisory teams, regiment and brigade in Korea and El Salvador and battalion in Vietnam, the teams were authorized only two to five personnel. As previously noted, personnel shortages were common. In Vietnam, district and province teams involved in pacification found that the variety of their responsibilities and the expanse of their territory stretched their capabilities. While these teams seem small for their tasks, Sir Robert Thompson, a British counterinsurgency expert in Malaya, believes in the "need on the military side to keep the presence of foreign military advisors to the minimum. If things are not going right, it is most unlikely that the solution will be found merely by increasing the quantity of advisors. This is liable to be counterproductive and can reach the point at which advice begins to revolve on a closed circuit."[2] In the advisory and in the counterinsurgency business more was not always better. Often more was less effective, particularly when it was more of the wrong stuff provided by advisors who were frequently rotating in and out of positions because of short-tour lengths and assignment policies.

Advisor roles and duties evolved, particularly in Vietnam and El Salvador. For combat unit advisors—even for KMAG—training, teaching, coaching, liaison, observing, tactical advising, and providing combat support were relatively straightforward, basic military tasks complicated by language, cultural differences, and institutional barriers. After 1965, MACV unit advisors focused primarily on providing American combat support assets, liaison with US units, and reporting on the status of their units. By the end of the El Salvador experience, a major duty for OPATT members was monitoring units for human rights abuses. For pacification advisors in Vietnam, civil-military duties in a counterinsurgency environment proved a complex, difficult task—something beyond their experience and expertise.

Who Advisors Were: Selection

If one thought that the small number of advisors at each level and the range of tasks required would mean a special selection process to identify those suitable, that proved not to be the case. While the advisor requirements in El Salvador did emphasize personnel with Special Forces experience, prior service in the region, and Spanish language capability, the general overall attitude was that anyone who spoke Spanish and had served in the region could do advisory duty. There was no special screening before training, except for the limited number of PSA and DSA positions at the end of Vietnam. If someone met rank and branch-qualification requirements and was eligible for an overseas tour, then he was suited for advisory duty. If he volunteered, it was even better. Frequently, even the basic rank, MOS, and experience requirements were waived as shown by the presence of first lieutenant battalion advisors and MAT team chiefs. Advisory duty, even in El Salvador, was never top priority. As an OPATT chief and Special Forces officer noted, "There were . . . criteria for technical or logistical advisor positions, and yet there is no system within the military to measure an advisor's professional competence in required skills, other than the language requirement. This is especially true for officers."[3] Consequently, in each of our experiences, the quality and suitability of advisory personnel varied. A MILGROUP commander observed that in Vietnam "Advisors, who should be the first team, were not; the nonmilitary aspects of the conflict we were in had been singled out as not being part of our job description; and there was an overall assumption that US combat units will win the war primarily with firepower."[4]

How Advisors Prepared: Training and Orientation

Other than Vietnam, advisors did not receive any special training prior to their duty assignments. Training was not considered necessary for combat unit advisors in Korea or in El Salvador for those with Spanish language skills and some experience in the region. Even the training for Vietnam—

focused on language, cultural, and advisory than on military skills—was considered only an introduction to the challenges faced by advisors. The quantity of advisors needed and the time required to develop awareness, much less understanding, precluded much more. PSA and DSA, those involved most directly in pacification, received longer language and special training toward the end of the Vietnam experience. However, "in both the cases of Vietnam and El Salvador, military planners were not worried about attendance to advisor courses that dealt with culture, language, or irregular warfare. It was the (MOS) schools that took priority."[5]

Interestingly, it is almost impossible to find a complaint by any advisor in the three experiences surveyed who felt tactically, technically, or militarily unprepared for his duties—even for those duties above his rank. Evidently MOS-qualification combined with American self-confidence met the basic military requirements faced by most advisors. However, almost to a man, advisors felt compelled to talk about the demanding challenges posed by language, cultural differences, and host-nation institutional barriers. It was in these areas—at the heart of an advisor's effectiveness—that most felt inadequately prepared. Although it is true that "there has never been a training program of instruction (POI) to prepare military advisors for duty that all those with an interest might agree was comprehensive and complete," it seemed clear that topics that enhanced situational understanding were considered more critical than those dealing with military skills.[6]

Not only was training generally nonexistent or of limited duration and value, the actual in-country orientation of new advisory personnel on the situation and procedures, with the exception of CORDS advisory personnel during Vietnam, was haphazard. In many cases, particularly in Korea, it was common for the new advisor to report directly to his KMAG team, with no orientation or briefing, not to mention his lack of specialized training or language skills. Even in El Salvador with its 55-man limit, at least one OPATT chief read the required folders and went directly to his unit, and did not see the MILGROUP commander for months. Low priority meant too few folks to do too much work which meant routine things were not routine.

How Advisors Did: Strengths and Weaknesses

In *Seven Pillars of Wisdom*, T.E. Lawrence wrote, "I was sent to these Arabs as a stranger, unable to think their thoughts or subscribe their beliefs, but charged by duty to lead them forward and to develop to the highest any movement of theirs profitable to England in her war."[7] So it was with the American advisors in Korea, South Vietnam, and El Salvador. Although

Lawrence's skills and qualifications were superior to those of American advisors, American advisors gave their best effort despite arduous circumstances. In Korea, the result was considered a success. With KMAG and EUSA support, ROKA expanded, improved, and ultimately proved capable of conducting effective combat operations against the North Korean and Chinese communist forces. In South Vietnam, the effort was considered a failure. With MACV support, RVNAF expanded and improved, but ultimately proved incapable of conducting effective combat operations against the North Vietnamese. In El Salvador, the results were mixed. With MILGROUP support, ESAF expanded, improved, and fought the FMLN to a stalemate, but ultimately proved incapable of internal reform or of conducting effective counterinsurgency operations. Although factors beyond their control often proved decisive, the final results were definitely influenced by the work of American advisors.

In advisory and counterinsurgency efforts, Thomas Carlyle's warning that "nothing is more terrible than activity without insight" is particularly appropriate. As a former MILGROUP commander wrote, "the problem is, and has always been, to get the analysis right *before* prescribing cures."[8] Analysis requires situational understanding, not awareness. Even in peacetime, under normal conditions, situational understanding can prove fleeting. In wartime, for an advisor in a foreign country, it is almost impossible. At a minimum, an advisor needs to understand the local language, the local culture and values, the local military institutional ethos and how it works, his counterpart as a person in that foreign culture and constrained by that military institution, the local capabilities and limitations, and the specific local situation to comprehend what is going on around him and to preclude misunderstandings. Then, it may be possible to offer advice suitable to the situation; acceptable both to his counterpart and to his US superiors; and feasible given time, resources, and the capabilities and limitations of host nation forces.

Although difficult, it is imperative that advisors just as all military personnel receive appropriate training to master the skill sets necessary for mission accomplishment. A review of the American experiences in Korea, South Vietnam, and El Salvador indicates that advisors faced significant challenges and suffered major shortcomings.

Advisors Were Deaf

Lacking language skills, advisors were basically deaf. They did not understand what was being said around them. In Korea, advisors were totally dependent on their ROKA translators. In South Vietnam, even with some basic language training, advisors were heavily dependent on

110

their RVNAF translators. In El Salvador, where some language skill was required, few advisors were native speakers. Without language training, communication is impaired. Unfortunately, suggestions that "to surmount the language barrier the American advisor had to be an inventive teacher, combining enthusiasm and knowledge with patience and tact," missed the point.[9]

Advisors Were Partially Blind

Unable to understand or visualize what was going on around them, advisors were partially blind. Not understanding the local cultural issues, the host-nation military institutional norms and procedures, and the specifics of local conditions, most advisors frequently misunderstood important things. This undercut rapport and increased the frustration and strain between the advisor and his counterpart. Although "frequently the success of the advisor depended as much upon his behavior as upon his professional ability," many advisors were unaware of the implications of their actions and inactions.[10] At best, instruction and orientation may have made advisors aware of some of these issues, but without understanding not only the "what," but more importantly the "why" the locals did things as they did, it created an illusion of knowledge.

Advisors Worked in a Hostile Environment

Even with support from above, advisors worked in a foreign country where US goals, techniques, procedures, and doctrines—not to mention language, culture, institutional imperatives—were not those of the host nation. Although desirous of American training and resources, no host nation wanted to become a clone of the US military. They wanted to become more combat-effective forms of what they were. Sir Robert Thompson emphasized,

> It is essential, therefore, for the advisor to look at everything from the local point of view and not to expect that the provision of aid will do more than provide the very limited benefits for which it was intended. He cannot expect that the threatened country will either organize itself or conduct its affairs on the same lines, or in accordance with the same standards, as those of the supporting power. The real point here, which is all the advisor can hope for, is to get the local government to function effectively and at least to take the necessary action itself, even if it is done in its own traditional way.[11]

Advisors found it relatively simple to train basic technical and tactical skills, but almost impossible to make those deeper institutional changes without host nation support from above, as in Korea. And even there, in the end, ROKA remained uniquely Korean, despite its American organizational structure and equipment.

Advisors Worked with Indifferent Counterparts

Because of the hostile environment and the frequent turnover of advisors, most counterparts were resistant—indifferent at best if not actually hostile—to an advisor's attempt to establish rapport. Without personal and institutional reasons to bond, and given the short-term focus of advisors serving brief assignments, establishing rapport was a topic much easier discussed than accomplished. Just because an American advisor showed up full of enthusiasm with a myriad of projects to improve the unit immediately, it did not translate into the counterpart sharing that enthusiasm, immediate focus, nor myriad of projects, most probably of questionable utility. When lower-level advisors could provide their counterpart something of value, such as combat support assets as they did in Korean and South Vietnam or pacification assets as in South Vietnam, then the counterpart had a personal incentive to work more closely with his advisor.

Advisors Worked with Limited US Military Support

Advisory duty was never the primary focus in Korea, South Vietnam, or El Salvador. It was an important effort, but secondary nonetheless. As such it received less support, fewer resources, little guidance, and often outdated or inappropriate doctrine. During Vietnam and El Salvador, the US military struggled with counterinsurgency operations while trying to develop low intensity or counterinsurgency doctrine. Without a grasp of the role of combat units in a counterinsurgency, much less the complex civil-military implications, American advisors proved more comfortable with the theory and less adept at turning theory into practice for a host nation.

Advisors Did Not Stay Long

A 1-year tour as an advisor was often viewed as something to be endured or gotten through. However, a former OPATT chief reported, "Most ex-advisors report that advisor tours longer than 1 year are absolutely essential to a successful effort."[12] Counterparts and repeated studies supported the view that advisors did not stay in their jobs long enough to understand the situation or to develop the rapport with their counterpart necessary to be productive. Longer tours provided longer-term approaches and more incentive for advisors to master the language, culture, and other factors necessary for doing the job effectively.

112

Although most American advisors were not conscience of it, they brought US military approaches developed to optimize US military organizations, systems, doctrine, and equipment against a Soviet or conventional threat that were often not appropriate for solving the problems faced by the host nation, particularly in a counterinsurgency. In fact, many were counterproductive. They had not learned that an "advisor must abandon the idea that his way is always best, and try to fit in and listen rather than provide advice by the book."[13] A historian and long-time student of American culture noted:

> The apparently irresistible bias of the American military is to train other nations' military organizations as our clones. This typically includes the Americans insisting on the bureaucratic organization of other armies into divisions, dependent on technology and committed to enormous firepower. It was inevitable in Vietnam that we would train a local air force, supply the planes and bombs, establish military academies, institute annual performance reviews, and suffer daily disappoint that the Vietnamese seemed unable to do the job. The unanswered question was what was the job. Because we were unable to put our own counterinsurgency tactics into operations, we trained the South Vietnamese to be similarly incompetent. Counterinsurgency, political war, required discipline and clarity to avoid using artillery and bombs. The American[s] . . . would never relinquish . . . technological superiority.[14]

The American military tended to do what it knew best, whether appropriate or not. When confronted with lack of host nation progress, the solution often was to increase the effort. The American way seemed better, quicker, "our way." Often "can do," "make it happen," "get over it," and "just do it" became substitutes for thought and analysis, resulting in more of the same, often done better but without a different result. Cultural understanding of others begins with cultural self-knowledge. American capabilities and limitations need to be explicitly defined, just as those of the host nation.

Final Thoughts

The prognosis of a "long war" working with host nation forces and allies in a counterinsurgency environment means that the likelihood of advisory duty for many US military personnel is almost a certainty. With the growing number of American military advisory teams—Mobile Training Teams (MTT), Military Transition Team (MiTT), Border Transition Team (BTT), Regional Border Transition Team (RBTT), Special Police Transition Team (SPTT), Embedded Training Team (ETT), and Provincial Reconstruction Team (PRT)—one thing remains constant: the requirement to advise effectively.

Because there is not a lot of material for potential advisors to consult, suggestions for working with counterparts developed from T.E. Lawrence, Korea, and Vietnam have been included in appendixes A through G. Appendix H, "Points for Consideration," and Appendix I, "21 Recommended Practices in Working with Counterparts" are extracted from the October 2001 *Special Forces Advisor's Reference Handbook*, an exceptional source that focuses primarily on advisor-related issues. Three firsthand advisory experiences are currently available commercially in paperback: Martin J. Dockery, *Lost in Translation: VIETNAM: A Combat Advisor's Story*, an excellent account of an battalion advisor from 1962 to 1963; David Donovan, *Once A Warrior King: Memories of an Officer in Vietnam*, the account of a MAT team chief and district senior advisor from 1969 to 1970; and Stuart A. Herrington, *Stalking the Vietcong: Inside Operation Phoenix: A Personal Account*, an account of a District Intelligence Operations Coordinating Center advisor from 1971 to 1972. In addition to these, the bibliography offers other sources for exploring the topic in more depth.

Based on this survey of American advisors in Korea, Vietnam, and El Salvador, the following thoughts are offered for those involved in planning, training, and directing advisory work.

• Advisory duty is a complex and difficult job, even more so in a counterinsurgency environment. Working effectively with indigenous forces in a foreign country—alien culture, unknown language, strange ways, incomprehensible actions, different concepts of right and wrong, good and bad, appropriate and inappropriate—is probably the hardest military task. As such, it takes time and a long-term focus. There are no quick or easy fixes; however, this survey indicates that the advisory effort can be more effective.

• Careful selection and screening of advisory personnel is required. Not everybody can or should do advisory duty. Former advisors acknowledge this; studies reinforce it. This means "to have a valid set of selection criteria that works, the military has to formulate a hard set of required skills for advisor duty. It should . . . then test them to ensure some level of proficiency."[15] "Good Marines [and good soldiers] do not invariably make good advisors . . . [for many] lacked the patience to work with a culture that places little emphasis on qualities that we regard as . . . indispensable to military life. . . . The 'drill instructor' type of instruction is not generally effective in training indigenous soldiers."[16] Those soldiers considered the best and most experienced are not always well suited for advisory duty; often the normal approach is also not well suited.

- Training and educational programs for advisory personnel should focus on the knowledge and skills necessary for understanding their advisory environment and for developing useful advice. Host nation cultural, language, military institutional, interpersonal relations, situational, and general advisory skills are more difficult to understand and to master and are more critical to advising than technical, tactical, and military skills that are normally basic and generally well-understood by MOS-qualified personnel. An exception is counterinsurgency operations with their civil-military and pacification programs. Training should focus on understanding—not awareness and familiarization. Language training is a must for situational understanding and for communicating advice. If interpreters are used, they should have the language skills of a native speaker, they should be trained to understand things military, and they should work for the advisor, not the host nation.

- The advisory effort should focus on how host nation organizations, institutions, systems, capabilities, and limitations—not US organizations, systems, procedures, and equipment—can be harnessed to address the host nation problems. As one American counterinsurgency expert recently noted, "Creating reliable, dedicated local forces . . . is a task as difficult as 'eating soup with a knife.'" Yet it is the critical task. "Local forces have inherent advantages over outsiders . . . intelligence . . . don't need . . . translators . . . understand the tribal loyalties and family relationships . . . innate understanding of local patterns of behavior that is simply unattainable by foreigners."[17] To tap into those advantages, the advisor must resist the "US military solution." To overcome the temptation to do what he knows and does best, whether relevant or not to the situation, each advisor must accept that he is "bound by a unique set of organizational [US military] fetters. Only by understanding those bindings can he take action to make them less confining or crippling, but never can he hope to strike them, once and for all, from his wrists."[18] As advisors in Iraq recently learned, "simply training, equipping, and organizing is not enough. We cannot undo the influence and corruption that has existed for hundreds of years by sending soldiers to a school, calling them commando, and expecting them to execute. It just isn't that easy."[19]

- Longer, repetitive advisory tours increase the effectiveness of advisors. One-year or shorter assignments reduce the incentives to prepare properly for advisory duty and limit the time to develop rapport, to understand the situation, and to develop practical advice. As Sir Robert Thompson observed about US advisors in Vietnam, "However high the caliber and it was uniformly good, no great achievements in counterinsurgency are possible in such a short period. All that the individual can hope to do is to leave his post at the end of the year as he would like to find it. He cannot do more than prepare the ground for his successors."[20] Studies and comments by ex-advisors support this finding.

- Although leverage at lower advisory levels threatens rapport, leverage at the highest host nation and US military advisory group level can enhance rapport and effectiveness by establishing common standards and expectations for both advisors and their counterparts. When this happens, the acceptance of common roles and responsibilities for working toward a common goal becomes possible. The responsibility for working effectively together should not rest just on the American advisor. The host nation needs to step up and train its personnel to work with American advisors by focusing on the same sort of things that the advisors do—language, culture, capabilities, and limitations—to better understand things from the American military point of view. It is the responsibility of both the host nation and the senior US advisory team to create an environment in which an advisor can develop the rapport with his counterpart necessary to work effectively together. A former advisor suggested "a national level agreement . . . to spell out, in writing, what are the specific functions of the advisors, what they will do and how they will do them. . . . The function of the [senior US advisory team] is to ask the HN forces if the advisors are doing what the HN wants and ask the advisors if they are being listened to and productively employed."[21] This coincides with the view of Sir Robert Thompson who stressed the importance of a treaty or formal written agreement. Without one, he stated, "Loose ends have a tendency to flap, and will flap their hardest at the end of the line where the policy has to be implemented on the ground," exactly where advisors and their counterparts are.[22] Much harder to accomplish than to propose, this approach clarifies functions, expectations, and feedback—significant advantages for a successful advisory effort. Unfortunately in our three cases, only when both sides in Korea recognized that they faced dire circumstances did a common approach happen. And even then, significant challenges—cultural, linguistic, institutional—remained.

- Rapport—that personal relationship of trust and confidence in one another's competence, motivation, and honesty—is always critical at the lower advisory levels. It does not happen automatically; it takes time and effort by both the counterpart and his advisor to develop. However, with a common understanding of expectations and roles articulated from above to both the counterpart and to his advisor, rapport, when it develops, truly becomes a force multiplier. Because the problems are complex and differences real, misunderstandings will still occur. Things will still go wrong. But at the level where things get done, rapport provides the grease that keeps things moving with minimal friction.

A Final Question

The observations of this brief survey, when combined with current advisory activities in Afghanistan and Iraq, point to a final question. Given the anticipated demands of the "long war" on US military personnel and increased advisory duties in countries facing counterinsurgency threats that require increased cultural and linguistic competence for working effectively with indigenous forces, does the US military need a single agency or proponent that has responsibility for advisory issues—concepts, requirements, doctrine, training, selection, planning, and operations? Perhaps it is now time to move beyond the ad hoc, make-it-up-as-you-go approach of the past to a more systematic and experience-based method. Only time will tell if the current Joint Center for International Security Force Assistance (JCISFA), Foreign Security Force Training Center (FSFTC), and Military Transition Team (MiTT) efforts are adequate.

The insights of T.E. Lawrence are as appropriate here as they were in the introduction of this survey. In a 26 April 1933 letter to B.H. Liddell Hart commenting on the draft of a military biography that Hart was writing, Lawrence wrote:

> Do make it clear that generalship, at least in my case, came of understanding, of hard study, and brain-work and concentration. Had it come easy to me I should have not done it so well. If your book could persuade some of our new soldiers to read and mark and learn things outside drill manuals and tactical diagrams, it would do a good work. I feel a fundamental, crippling, incuriousness about our officers. Too much body and too little head. The perfect general would know everything in heaven and earth. So please, if you see me that way and agree with me, do use me as a text to preach for more study of books and history, a greater seriousness in military art. With 2,000 years of examples behind us we have no excuse, when fighting, for not fighting well.[23]

When Lawrence used the term generalship, he was describing his role as advisor to Emir Feisal during the Arab Revolt. To paraphrase his final sentence: since the beginning of World War II, with over 65 years of American examples behind us, we have no excuse, when advising, for not advising well.

Notes

1. Walter G. Hermes, "Survey of the Development of the Role of the U.S. Army Military Advisor" (Washington, DC: Office of the Chief of Military History, Historical Report, 1965), 80.

2. Robert Thompson, *Defeating Communist Insurgency: The Lessons of Malaya and Vietnam* (New York, NY: Frederick A. Praeger, 1966), 165.

3. Mark A. Meoni, "The Advisor: From Vietnam to El Salvador" (Master of Military Art and Science Thesis, US Army Command and General Staff College, 1992), 138.

4. John D. Waghelstein, "What's Wrong in Iraq? Or Ruminations of a Pachyderm," *Military Review* (January–February 2006), 114.

5. Meoni, 142.

6. Ibid., 184.

7. T.E. Lawrence, *Seven Pillars of Wisdom: A Triumph* (New York, NY: Anchor Books, 1991), 30.

8. Waghelstein, 117.

9. Hermes, 82.

10. Ibid., 82–83.

11. Thompson, 161.

12. Meoni, 218.

13. Ibid., 155.

14. Loren Baritz, *Backfire: Vietnam—The Myths That Made Us Fight, The Illusions That Helped Us Lose, The Legacy That Haunts Us Today* (New York, NY: Ballentine Books, 1985), 243.

15. Meoni, 139.

16. Andrew R. Milburn and Mark C. Lombard, "Marine Foreign Military Advisors: The Road Ahead," *Marine Corps Gazette* (April 2006), 62–63.

17. John A. Nagl, *Learning to Eat Soup with a Knife: Counterinsurgency Lessons from Malaya and Vietnam* (Chicago, IL: University of Chicago Press, 2005), xiv.

18. Eliot A. Cohen and John Gooch, *Military Misfortunes: The Anatomy of Failure in War* (New York, NY: Free Press, 2006), 246.

19. David H. Marshall, "Training Iraqi Forces," *Marine Corps Gazette* (April 2006), 60.

20. Thompson, 165.

21. "After Action Report, 2d MILZONE OPATT Chief, Apr 91–Apr 92," Memorandum for Commander, MILGROUP, 18 April 1992.

22. Thompson, 158.

23. Jeremy Wilson, *Lawrence of Arabia: The Authorized Biography of T.E. Lawrence* (New York: Antheneum, 1990), 908.

About the Author

Robert D. Ramsey III retired from the US Army in 1993 after 24 years of service as an Infantry officer that included tours in Vietnam, Korea, and the Sinai. He earned an M.A. in history from Rice University. Mr. Ramsey taught military history 3 years at the United States Military Academy and 6 years at the US Army Command and General Staff College.

Glossary

AAR	after action review
ALAT	Army Language Aptitude Test
AR	Army regulation
ARVN	Army of Vietnam
BCAT	Battalion Combat Assistance Team
BTT	Border Transition Team
CA	Civic Action; California
CAS	close air support
CAT	combat assistance team
CCF	Chinese Communist Forces
CCP	Combined Campaign Plan (Vietnam)
CG	Civil Guard (Vietnam)
CGSC	Command and General Staff College
CIA	Central Intelligence Agency
CIDG	Civilian Irregular Defense Group
CMO	civil military operations
CO	commanding officer
COIN	counterinsurgency
CONUS	continental United States
CORDS	Civil Operations and Revolutionary Development Support
CP	command post
CSA	Chief of Staff of the Army
DA	Department of the Army
DCAT	Division Combat Assistance Team
DCSPER	Deputy Chief of Staff for Personnel
DIOCC	District Intelligence Operations and Coordination Center
div	division
DOD	Department of Defense
DOS	Department of State
DPSA	Deputy Province Senior Advisor
DSA	District Senior Advisor
e.g.	for example
ESAF	El Salvador Armed Forces
etc.	and so forth
ETT	Embedded Training Team
EUSA	Eighth US Army (Korea)
FAO	Foreign Area Officer
FASP	Foreign Area Specialist Program
FM	frequency modulation; field manual
FMLN	*Farabundo Martí para Liberación Nacional* movement (El Salvador)
FSF	Foreign Security Force
FSFTC	Foreign Security Force Training Center

G1	Assistant Chief of Staff (Personnel)
G2	Assistant Chief of Staff (Intelligence)
G3	Assistant Chief of Staff (Operations and Plans)
G4	Assistant Chief of Staff (Logistics)
G5	Assistant Chief of Staff (Civil Affairs)
GOES	Government of El Salvador
GVN	Government of Vietnam
GWOT	Global War on Terrorism
HC	host country
HN	host nation
HQ	headquarters
i.e.	that is
IMA	Institute for Military Assistance
inf	infantry
JCISFA	Joint Center for International Security Force Assistance
JCS	Joint Chiefs of Staff
JSAT	Joint Security Assistance Training
KISSSS	Keep it simple, sustainable, small, and Salvadorian
KMAG	United States Military Advisory Group to the Republic of Korea
LIC	low intensity conflict
LTG	lieutenant general
MAAG	Military Assistance Advisory Group
MAAG-V	Military Assistance Advisory Group, Vietnam
MACV	Military Assistance Command, Vietnam
MAG	Military Advisory Group
MAI	Military Assistance Institute
MALT	Mobile Advisory Logistics Team
MAOP	Military Assistance Officer Program
MAPA	Military Assistance Program Advisor
MASA	Military Assistance Security Advisor
MAT	Mobile Advisory Team
MATA	Military Assistance Training and Advisory (course)
MD	Maryland
MI	military intelligence
MILGROUP	Military Group (El Salvador)
MILZONE	military zone
MiTT	Military Transition Teams
MOS	Military Occupational Specialty
MTT	Mobile Training Team
NC	North Carolina
NCO	noncommissioned officer
NGO	non-governmental organization
NVA	North Vietnamese Army
ODCSPER	Office of the Deputy Chief of Staff for Personnel

OP	Occasional Paper
OPATT	Operations, Plans and Training Team (El Salvador)
OSO	Overseas Security Operations
PF	Popular Forces (initially Self-Defense Corps) (Vietnam)
PIOCC	Province Intelligence Operations and Coordination Center
PMAG	Provisional Military Advisory Group (Korea)
POI	program of instruction
PRT	Provisional Reconstruction Team
PSA	Province Senior Advisor
PSDF	People's Self-Defense Force
PSYOPs	psychological operations
QDR	Quadrennial Defense Review
R&R	rest and relaxation
R2	Limited Working Proficiency—Reading
R3	General Professional Proficiency—Reading
RBTT	Regional Border Transition Team
RCAT	Regiment Combat Assistance Team
RF	Regional Forces (initially Civil Guard) (Vietnam)
ROK	Republic of Korea
ROKA	Republic of Korea Army
RVN	Republic of Vietnam
RVNAF	Republic of Vietnam Armed Forces
S2	Limited Working Proficiency—Speaking
S3	General Professional Proficiency—Speaking
S5	Civil Affairs Officer
SATTOC	Security Assistance Team Training and Orientation Course
SDC	Self Defense Corps (Vietnam)
SE	southeast
SF	Special Forces
SOI	surety operational inspection
SOUTHCOM	US Southern Command
SPTT	Special Police Transition Team
SSI	selection source information
TDY	temporary duty
TERM	Temporary Equipment Recovery Mission (Vietnam)
TTP	tactics, techniques, and procedures
TZ	tactical zone
US	United States
USAF	US Air Force
USAID	US Agency for International Development
USARV	US Army, Vietnam
USIA	US Information Agency
USMC	US Marine Corps
USOM	US Operations Mission

VA	Virginia
VC	Viet Cong
VN	Vietnamese; Vietnam

Bibliography

Official US Army and US Marine Corps Studies and Reports

Appleman, Roy E. *South to the Naktong, North to the Yalu (June–November 1950)*. Washington, DC: US Army Office of the Chief of Military History, 1961.

Clarke, Jeffrey J. *Advice and Support: The Final Years, 1965–1973*. Washington, DC: US Army Center of Military History, 1988.

Collins, James L., Jr. *The Development and Training of the South Vietnamese Army, 1950–1972*. Washington, DC: Department of the Army, 1975.

Hermes, Walter G. "Survey of the Development of the Role of the U.S. Army Military Advisor." Washington, DC: Office of the Chief of Military History, Historical Report, 1965.

_____. *Truce Tent and Fighting Front*. Washington, DC: US Army Office of the Chief of Military History, 1966.

Kelly, Francis J. *U.S. Army Special Forces, 1961–1971*. Washington, DC: Department of the Army, 1973.

Mossman, Billy C. *Ebb and Flow: November 1950–July 1951*. Washington, DC: US Army Center of Military History, 1990.

Ott, David Ewing. *Field Artillery, 1954–1973*. Washington, DC: Headquarters, Department of the Army, 1975.

Sawyer, Robert K. *Military Advisors in Korea: KMAG in Peace and War*. Washington, DC: US Army Center for Military History, 1962.

Spector, Ronald H. *Advice and Support: The Early Years, 1941–1960*. Washington, DC: US Army Center of Military History, 1983.

Truong, Ngo Quang. *RVNAF and US Operational Cooperation and Coordination*. Washington, DC: US Army Center of Military History, 1980.

Vien, Cao Van, Ngo Quang Truong, Dong Van Khuyen, Nguyen Duy Hinh, Tran Ding Tho, Hoang Ngo Lung, and Chu Xuan Vien. *The U.S. Adviser*. Washington, DC: US Army Center of Military History, 1980.

Whitlow, Robert H. *U.S. Marines in Vietnam: The Advisory & Combat Assistance Era, 1954–1964*. Washington, DC: US Marine Corps History and Museums Division, 1977.

Field Manuals, Pamphlets, and Documents

"After Action Report, 2d MILZONE OPATT Chief, Apr 91–Apr 02." Memorandum for Commander, USMILGP, 18 April 1992.

Department of the Army. Field Manual (FM) 31–73, *Advisor Handbook for Stability Operations*. Washington, DC: Headquarters, Department of the Army, October 1967.

Department of State. "Vietnam Training Center District Operations Course," Foreign Service Institute, 1 October 1969. Available as document # 1070217055 in the Glenn Helm Collection, Texas Tech University Vietnam Archive Oral History Project.

Headquarters, Delta Regional Assistance Command, Vietnam. "Senior Officer Debriefing Report of Major General John H. Cushman, RCS CSFOR– 74," 14 January 1972.

Headquarters, United States Military Assistance Command, Vietnam. "Command History 1964," 15 October 1965.

_____. "Command History 1965," 20 April 1966.

_____. "Command History 1966," 30 June 1967.

_____. "Command History 1967," 3 vols, 11 September 1968.

_____. "Command History 1968," 2 vols, 30 April 1969.

_____. "Command History 1969," 2 vols, 30 April 1970.

_____. "Command History 1970," 4 vols, 19 April 1971.

_____. "Command History 1971," 2 vols, 25 April 1972.

_____. "Command History 1972–1973," 2 vols, 15 July 1973.

_____. *Handbook for Military Support of Pacification*, February 1968.

_____. *rf / pf Advisor Handbook*, January 1971.

US Army Section, Military Advisory and Assistance Group, Vietnam. *Book of Instructions for U.S. Military Advisors to 42 Tactical Zone III Corps South Vietnam*, 1962.

US Army Special Warfare School. *MATA Handbook for Vietnam*. Fort Bragg, NC: US Army Special Warfare School, January 1966.

_____. "Program of Instruction for Military Assistance Training Advisor Course (MATA)." Fort Bragg, NC: US Army Special Warfare School, April 1962.

_____. "Program of Instruction for Project 404." Fort Bragg, NC: US Army Special Warfare School, March 1969.

_____. *Special Forces Advisor's Reference Book*. Fort Bragg, NC: Research Planning, Inc., October 2001.

Directives and Regulations

Department of Defense. *Quadrennial Defense Review Report*. Washington, DC: Department of Defense, 6 February 2006.

Department of the Army. Army Regulation (AR) 12–15, *Joint Security Assistance Training (JSAT)*. Washington, DC: Headquarters, Department of the Army, 5 June 2002.

_____. AR 614–134, *Military Assistance Officer Program (MAOP)*. Washington, DC: Headquarters, Department of the Army, 20 June 1971.

Special Studies and Reports

BDM Corporation. *A Study of Strategic Lessons Learned in Vietnam*. Executive Summary and 8 vols., April 1980.

Childress, Michael. *The Effectiveness of U.S. Training Efforts in Internal Defense and Development: The Cases of El Salvador and Honduras*. Santa Monica, CA: RAND, 1995.

Draft Copy THE KMAG ADVISOR Circulated for Critical Reading by Selected Reviewers. Chevy Chase, MD: The Johns Hopkins University Operations Research Office, nd.

Erickson, Edwin E. and Herbert H. Vreeland, 3rd. *Operational and Training Requirements of the Military Assistance Officer*. McLean, VA: Human Sciences Research, Inc., May 1971.

Fiks, Alfred I. and John W. McCrary. *Some Language Aspects of the U.S. Advisory Role in South Vietnam*. Alexandria, VA: The George Washington University Human Resources Research Office, November 1963.

Foster, Robert J. *Examples of Cross-Cultural Problems Encountered by Americans Working Overseas: An Instructor's Handbook*. The George Washington University Human Resources Research Office, May 1965.

Guthrie, George M. *Conflicts of Culture and the Military Advisor*. Institute for Defense Analyses Research and Engineering Support Division, November 1966.

Hausrath, Alfred H. *The KMAG Advisor: Roles and Problems of the Military Advisor in Developing an Indigenous Army for Combat Operations in Korea*. Chevy Chase, MD: The Johns Hopkins University Operations Research Office, February 1957.

Hickey, Gerald C. *The American Military Advisor and His Foreign Counterpart: The Case of Vietnam*. Santa Monica, CA: RAND Corporation, March 1965.

Ridgway, Matthew B. "How the Korean Army Improved." Interview by Major Caulfield, October 1969. In *A Systems Analysis View of the Vietnam War: 1965–1972*. vol 7. *Republic of Vietnam Armed Forces (RVNAF)*. BDM Corporation, April 1980, 44–63.

Schwarz, Benjamin C. *American Counterinsurgency Doctrine and El Salvador: The Frustrations of Reform and the Illusions of Nation Building*. Santa Monica, CA: RAND Corporation, 1991.

Unpublished Dissertations, Theses, and Papers

Dawkins, Peter M. "The United States Army and the 'Other' War in Vietnam: A Study of the Complexity of Implementing Organizational Change." Ph.D. diss., Princeton University, 1979.

McCoy, Don S. "Military Assistance Command Unit Advisor Personal Characteristics, Training, and Techniques." Student Thesis, US Army War College, 12 February 1970.

Meoni, Mark A. "The Advisor: From Vietnam to El Salvador." Master of Military Art and Science Thesis, US Army Command and General Staff College, 1992.

Published Articles and Papers

Bailey, Cecil E. "OPATT: The U.S. Army SF Advisers in El Salvador." *Special Warfare*, December 2004, 18–29.

Brown, John S. "The Vietnam Advisory Effort." *Army*, March 2006, 94–95.

Castrillo, Victor J. "Contributions, Shortcomings, and Lessons Learned from U.S. MI Training/Advising in El Salvador." *Military Intelligence Professional Bulletin*, October–December 1993, 39–43.

Cole, Jeffrey U. "Assisting El Salvador." *Proceedings*, November 1989, 60–69.

Collins, James F. "The Individual Side of Counter-Insurgency." *Army,* August 1962, 28–32.

Dacus, David M. "So Now You're an Advisor." *Infantry,* May–June 1971, 32–35.

Denno, Bryce F. "Advisor and Counterpart." *Army,* July 1965, 25–30.

Hackworth, David H. "Our Advisors Must Pass the Ball." *Army,* May 1971, 61–62.

Hillman, Rolfe L., Jr. "Eclipse of the Advisor." *Army,* November 1966, 60–67.

Hudlin, Irving C. "Advising the Advisor." *Military Review,* November 1965, 94–96.

Jones, Richard A. "Nationbuilder: Soldier of the Sixties." *Military Review,* January 1965, 63–67.

Manwaring, Max G. and Court Prisk. "A Strategic View of Insurgencies: Insights from El Salvador." McNair Paper No. 8. Washington, DC: Institute for National Strategic Studies, National Defense University Press, May 1990.

Marshall, David H. "Training Iraqi Forces." *Marine Corps Gazette,* April 2006, 60.

Mendez, Marcos R. "The Role of an MI Advisor in El Salvador." *Military Intelligence Professional Bulletin,* October–December 1993, 28–30.

Milburn, Andrew R. and Mark C. Lombard. "Marine Foreign Military Advisors: The Road Ahead." *Marine Corps Gazette,* April 2006, 62–63.

Montgomery, Tommie Sue. "Fighting Guerrillas: The United States and Low-Intensity Conflict in El Salvador." *New Political Science,* Fall/Winter 1990, 21–53.

Pritchell, James H. "Subsector Advisor in Cai Lay District." *Army,* July 1966, 52–58.

Ray, James F. "The District Advisor." *Military Review,* May 1965, 3–8.

Rosello, Victor M. "Lessons from El Salvador." *Parameters,* Winter 1993–94, 100–108.

_____. "Reflections on El Salvador." *Military Intelligence Professional Bulletin,* October–December 1993, 26–27.

Shelton, David L. "Some Advice for the Prospective Advisor." *Marine Corps Gazette,* October 1991, 55–57.

Spector, Ronald H. "The First Vietnamization: U.S. Advisors in Vietnam, 1956–1960." In *The American Military and the Far East: Proceedings of the Ninth Military History Symposium, United States Air Force Academy, 1–3 October 1980,* edited by Joe C. Dixon, 109–115. Washington, DC: US Government Printing Office, 1980.

Steward, Edward C. "American Advisors Overseas." *Military Review,* February 1965, 3–9.

Waghelstein, John D. "Post-Vietnam Counterinsurgency Doctrine." *Military Review,* May 1985, 42–49.

_____. "Ruminations of a Pachyderm or What I Learned in the Counter-insurgency Business." *Small Wars and Insurgencies,* Winter 1994, 360–378.

_____. "What's Wrong in Iraq? Or Ruminations of a Pachyderm." *Military Review*, January–February 2006, 112–117.

Waters, Roland B. "The Marine Officer and a Little War." *Marine Corps Gazette*, December 1986, 54–59.

Books

Appleman, Roy E. *Disaster in Korea: The Chinese Confront MacArthur*. College Station, TX: Texas A&M Press, 1989.

_____. *Ridgway Duels for Korea*. College Station, TX: Texas A&M University Press, 1990.

Bacevich, Andrew J., James D. Hallums, Richard H. White, and Thomas F. Young. *American Military Policy in Small Wars: The Case of El Salvador*. New York, NY: Pergamon-Brassey's, 1988.

Baritz, Loren. *Backfire: Vietnam—The Myths That Made Us Fight, The Illusions That Helped Us Lose, The Legacy That Haunts Us Today*. New York, NY: Ballentine Books, 1985.

Blair, Clay. *The Forgotten War: America in Korea, 1950–1953*. New York, NY: Times Books, 1987.

Cohen, Eliot A. and John Gooch. *Military Misfortunes: The Anatomy of Failure in War*. New York, NY: Free Press, 2006.

Cook, John. *The Advisor*. New York, NY: Bantam, 1973.

Dockery, Martin J. *Lost in Translation: VIETNAM: A Combat Advisor's Story*. New York, NY: Ballantine, 2003.

Donovan, David. *Once a Warrior King: Memories of an Officer in Vietnam*. New York, NY: Ballantine, 1985.

Frehrenbach, T.R. *This Kind of War: A Study in Preparedness*. New York, NY: The Macmillan Company, 1963.

Herrington, Stuart A. *Silence Was a Weapon: The Vietnam War in the Villages*. Novato, CA: Presidio Press, 1982.

Kinnard, Douglas. *The War Managers*. Hanover, NH: University Press of New England, 1977.

Lawrence, T.E. *Seven Pillars of Wisdom: A Triumph*. New York, NY: Anchor Books, 1991.

Manwaring, Max G. and Court Prisk, eds. *El Salvador at War: An Oral History of Conflict from the 1979 Insurrection to the Present*. Washington, DC: National Defense University Press, 1988.

Miller, John G. *The Co-Vans: U.S. Marine Advisors in Vietnam*. Annapolis, MD: Naval Institute Press, 2000.

Nagl, John A. *Learning to Eat Soup with a Knife: Counterinsurgency Lessons from Malaya and Vietnam*. Chicago, IL: University of Chicago Press, 2005.

Paik Sun Yup. *From Pusan to Panmunjom*. Dulles, VA: Brassey's, 1999.

Palmer, Bruce, Jr. *The 25 Year War: America's Military Role in Vietnam*. Lexington, KY: University of Kentucky Press, 1984.

Ridgway, Matthew B. *The Korean War*. New York, NY: Da Capo Press, 1967.

Thompson, Robert. *Defeating Communist Insurgency: The Lessons of Malaya and Vietnam*. New York, NY: Frederick A. Praeger, 1966.

Westmoreland, William C. *A Soldier Reports*. New York, NY: Dell, 1980.

Willbanks, James H. *The Battle of An Loc*. Bloomington, IN: Indiana University Press, 2005.

Wilson, Jeremy. *Lawrence of Arabia: The Authorized Biography of T.E. Lawrence*. New York: Antheneum, 1990.

Internet Sources

Barron, Robert. Interview by Stephen Maxner, 21 April 2001. Texas Tech University Vietnam Archive Oral History Project. Transcript online; available at http://star.vietnam.ttu.edu/starweb/vva/servlet.starweb?path=vva/oral history.web&id=newweboh&pass=&search1=WEBOK%3DYES&format=format.

Colby, William E. "U.S. Assistance Programs in Vietnam: Hearings Before a House of Representatives Subcommittee of the Committee on Government Operations 19 July 1971." Document online; available at http://homepage.ntlworld.com/jksonc/docs/phoenix-hcgo-19710719.html.

Department of the Army. "Department of the Army Historical Summary: FY 1971." Document online: available at http://www.army.mil/cmh/books/DAHSUM/1971/chIV.htm.

————. "Department of the Army Historical Summary: FY 1972." Document online: available at http://www.army.mil/cmh/books/DAHSUM/1972/ch02.htm.

Haseman, John. Interview by Stephen Maxner, 21 June 2000. Texas Tech University Vietnam Archive Oral History Project. Transcript online; available at http://star.vietnam.ttu.edu/starweb/vva/servlet.starweb?path=vva/oralhistory.web&id=newweboh&pass=&search1=WEBOK%3DYES&format=format.

Jones, Eddie. Interview by Stephen Maxner, 13 January 2003. Texas Tech University Vietnam Archive Oral History Project. Transcript online; available at http://star.vietnam.ttu.edu/starweb/vva/servlet.starweb?path=vva/oralhistory.web&id=newweboh&pass=&search1=WEBOK%3DYES&format=format.

Kinzer, Joseph W. "End-of-Tour Interview" by Charles C. Pritchett, 11 June 1968. US Army Center of Military History. Oral History Interview VNIT 101 online; available at http://www.army.mil/cmh-pg/documents/vietnam/vnit/vnit0101.htm.

Lawrence, T.E. "The 27 Articles of T.E. Lawrence." In *The Arab Bulletin*, 20 August 1917. Document online; available at http://www.lib.byu.edu/~rdh/wwi/1917/27arts.html.

Moriarty, William. Interview by Shelby Sears, 24 June 2004. Virginia Military Institute John A. Adams '71 Center for Military History and Strategic Analysis Cold War Oral History Project. Transcript online; available at http://www.vmi.edu/archives/Adams_Center/MoriartyB/MoriartyB_ intro.asp.

Ray, Ronald D. Interview by Shelby Sears, 21 June 2004. Virginia Military Institute John A. Adams '71 Center for Military History and Strategic Analysis Cold War Oral History Project. Transcript online; available at http://www.vmi.edu/archives/Adams_Center/RayR/RayR_intro.asp.

Shelton, William. Interview by Stephen Maxner, 5 June 2000. Texas Tech University Vietnam Archive Oral History Project. Transcript online; available at http://star.vietnam.ttu.edu/starweb/vva/servlet. starweb?path=vva/oralhistory.web&id=newweboh&pass=&search1= WEBOK%3DYES&format=format.

Woerner, Fred E. "Report of the El Salvador Military Strategy Assistance Team (DRAFT)." Document online; available at http://www.gwu.edu/ %7ensarchiv/nsa/DOCUMENT/930325.htm.

Wright, Jackie V. Interview by Dr. Richard Verrone, 12 December 2002. Texas Tech University Vietnam Archive Oral History Project. Transcript online; available at http://star.vietnam.ttu.edu/starweb/vva/servlet. starweb?path=vva/oralhistory.web&id=newweboh&pass=&search1=W EBOK%3DYES&format=format.

Zinni, Anthony C. Interview by Shelby Sears, 29 June 2004. Virginia Military Institute John A. Adams '71 Center for Military History and Strategic Analysis Cold War Oral History Project. Transcript online; available at http://www.vmi.edu/archives/archivecoldwar/Details.asp? ID=9&rform=search.

Appendix A

"Twenty-Seven Articles,"* 1917

The following notes have been expressed in commandment form for greater clarity and to save words. They are, however, only my personal conclusions, arrived at gradually while I worked in the Hejaz and now put on paper as stalking horses for beginners in the Arab armies. They are meant to apply only to Bedu; townspeople or Syrians require totally different treatment. They are of course not suitable to any other person's need, or applicable unchanged in any particular situation. Handling Hejaz Arabs is an art, not a science, with exceptions and no obvious rules. At the same time we have a great chance there; the Sherif trusts us, and has given us the position (toward his Government) which the Germans wanted to win in Turkey. If we are tactful, we can at once retain his goodwill and carry out our job, but to succeed we have got to put into it all the interest and skill we possess.

1.   Go easy for the first few weeks. A bad start is difficult to atone for, and the Arabs form their judgments on externals that we ignore. When you have reached the inner circle in a tribe, you can do as you please with yourself and them.

2.   Learn all you can about your Ashraf and Bedu. Get to know their families, clans and tribes, friends and enemies, wells, hills and roads. Do all this by listening and by indirect inquiry. Do not ask questions. Get to speak their dialect of Arabic, not yours. Until you can understand their allusions, avoid getting deep into conversation or you will drop bricks. Be a little stiff at first.

3.   In matters of business deal only with the commander of the army, column, or party in which you serve. Never give orders to anyone at all, and reserve your directions or advice for the C.O., however great the temptation (for efficiency's sake) of dealing with his underlings. Your place is advisory, and your advice is due to the commander alone. Let him see that this is your conception of your duty, and that his is to be the sole executive of your joint plans.

4.   Win and keep the confidence of your leader. Strengthen his prestige at your expense before others when you can. Never refuse or quash schemes he may put forward; but ensure that they are put forward in the first

*T.E. Lawrence, "Twenty-Seven Articles," *The Arab Bulletin*, 27 August 1917 [online]; available at http://www.lib.byu.edu/~rdh/wwi/1917/27arts.html.

instance privately to you. Always approve them, and after praise modify them insensibly, causing the suggestions to come from him, until they are in accord with your own opinion. When you attain this point, hold him to it, keep a tight grip of his ideas, and push them forward as firmly as possibly, but secretly, so that to one but himself (and he not too clearly) is aware of your pressure.

5. Remain in touch with your leader as constantly and unobtrusively as you can. Live with him, that at meal times and at audiences you may be naturally with him in his tent. Formal visits to give advice are not so good as the constant dropping of ideas in casual talk. When stranger sheikhs come in for the first time to swear allegiance and offer service, clear out of the tent. If their first impression is of foreigners in the confidence of the Sherif, it will do the Arab cause much harm.

6. Be shy of too close relations with the subordinates of the expedition. Continual intercourse with them will make it impossible for you to avoid going behind or beyond the instructions that the Arab C.O. has given them on your advice, and in so disclosing the weakness of his position you altogether destroy your own.

7. Treat the sub-chiefs of your force quite easily and lightly. In this way you hold yourself above their level. Treat the leader, if a Sherif, with respect. He will return your manner and you and he will then be alike, and above the rest. Precedence is a serious matter among the Arabs, and you must attain it.

8. Your ideal position is when you are present and not noticed. Do not be too intimate, too prominent, or too earnest. Avoid being identified too long or too often with any tribal sheikh, even if C.O. of the expedition. To do your work you must be above jealousies, and you lose prestige if you are associated with a tribe or clan, and its inevitable feuds. Sherifs are above all blood-feuds and local rivalries, and form the only principle of unity among the Arabs. Let your name therefore be coupled always with a Sherif's, and share his attitude toward the tribes. When the moment comes for action put yourself publicly under his orders. The Bedu will then follow suit.

9. Magnify and develop the growing conception of the Sherifs as the natural aristocracy of the Arabs. Intertribal jealousies make it impossible for any sheikh to attain a commanding position, and the only hope of union in nomad Arabs is that the Ashraf be universally acknowledged as the ruling class. Sherifs are half-townsmen, half-nomad, in manner and life, and have the instinct of command. Mere merit and money would be insufficient to obtain such recognition; but the Arab reverence for pedigree

and the Prophet gives hope for the ultimate success of the Ashraf.

10. Call your Sherif "Sidi" in public and in private. Call other people by their ordinary names, without title. In intimate conversation call a Sheikh "Abu Annad," "Akhu Alia" or some similar by-name.

11. The foreigner and Christian is not a popular person in Arabia. However friendly and informal the treatment of yourself may be, remember always that your foundations are very sandy ones. Wave a Sherif in front of you like a banner and hide your own mind and person. If you succeed, you will have hundreds of miles of country and thousands of men under your orders, and for this it is worth bartering the outward show.

12. Cling tight to your sense of humor. You will need it every day. A dry irony is the most useful type, and repartee of a personal and not too broad character will double your influence with the chiefs. Reproof, if wrapped up in some smiling form, will carry further and last longer than the most violent speech. The power of mimicry or parody is valuable, but use it sparingly, for wit is more dignified than humor. Do not cause a laugh at a Sherif except among Sherifs.

13. Never lay hands on an Arab; you degrade yourself. You may think the resultant obvious increase of outward respect a gain to you, but what you have really done is to build a wall between you and their inner selves. It is difficult to keep quiet when everything is being done wrong, but the less you lose your temper the greater your advantage. Also then you will not go mad yourself.

14. While very difficult to drive, the Bedu are easy to lead, if: have the patience to bear with them. The less apparent your interferences the more your influence. They are willing to follow your advice and do what you wish, but they do not mean you or anyone else to be aware of that. It is only after the end of all annoyances that you find at bottom their real fund of goodwill.

15. Do not try to do too much with your own hands. Better the Arabs do it tolerably than that you do it perfectly. It is their war, and you are to help them, not to win it for them. Actually, also, under the very odd conditions of Arabia, your practical work will not be as good as, perhaps, you think it is.

16. If you can, without being too lavish, forestall presents to yourself. A well-placed gift is often most effective in winning over a suspicious sheikh. Never receive a present without giving a liberal return, but you may delay this return (while letting its ultimate certainty be known) if you require a particular service from the giver. Do not let them ask you for

things, since their greed will then make them look upon you only as a cow to milk.

17. Wear an Arab headcloth when with a tribe. Bedu have a malignant prejudice against the hat, and believe that our persistence in wearing it (due probably to British obstinacy of dictation) is founded on some immoral or irreligious principle. A thick headcloth forms a good protection against the sun, and if you wear a hat your best Arab friends will be ashamed of you in public.

18. Disguise is not advisable. Except in special areas, let it be clearly known that you are a British officer and a Christian. At the same time, if you can wear Arab kit when with the tribes, you will acquire their trust and intimacy to a degree impossible in uniform. It is, however, dangerous and difficult. They make no special allowances for you when you dress like them. Breaches of etiquette not charged against a foreigner are not condoned to you in Arab clothes. You will be like an actor in a foreign theatre, playing a part day and night for months, without rest, and for an anxious stake. Complete success, which is when the Arabs forget your strangeness and speak naturally before you, counting you as one of themselves, is perhaps only attainable in character: while half-success (all that most of us will strive for; the other costs too much) is easier to win in British things, and you yourself will last longer, physically and mentally, in the comfort that they mean. Also then the Turks will not hang you, when you are caught.

19. If you wear Arab things, wear the best. Clothes are significant among the tribes, and you must wear the appropriate, and appear at ease in them. Dress like a Sherif, if they agree to it.

20. If you wear Arab things at all, go the whole way. Leave your English friends and customs on the coast, and fall back on Arab habits entirely. It is possible, starting thus level with them, for the European to beat the Arabs at their own game, for we have stronger motives for our action, and put more heart into it than they. If you can surpass them, you have taken an immense stride toward complete success, but the strain of living and thinking in a foreign and half-understood language, the savage food, strange clothes, and stranger ways, with the complete loss of privacy and quiet, and the impossibility of ever relaxing your watchful imitation of the others for months on end, provide such an added stress to the ordinary difficulties of dealing with the Bedu, the climate, and the Turks, that this road should not be chosen without serious thought.

21. Religious discussions will be frequent. Say what you like about your own side, and avoid criticism of theirs, unless you know that the point is

external, when you may score heavily by proving it so. With the Bedu, Islam is so all-pervading an element that there is little religiosity, little fervour, and no regard for externals. Do not think from their conduct that they are careless. Their conviction of the truth of their faith, and its share in every act and thought and principle of their daily life is so intimate and intense as to be unconscious, unless roused by opposition. Their religion is as much a part of nature to them as is sleep or food.

22. Do not try to trade on what you know of fighting. The Hejaz confounds ordinary tactics. Learn the Bedu principles of war as thoroughly and as quickly as you can, for till you know them your advice will be no good to the Sherif. Unnumbered generations of tribal raids have taught them more about some parts of the business than we will ever know. In familiar conditions they fight well, but strange events cause panic. Keep your unit small. Their raiding parties are usually from one hundred to two hundred men, and if you take a crowd they only get confused. Also their sheikhs, while admirable company commanders, are too "set" to learn to handle the equivalents of battalions or regiments. Don't attempt unusual things, unless they appeal to the sporting instinct Bedu have so strongly, unless success is obvious. If the objective is a good one (booty) they will attack like fiends, they are splendid scouts, their mobility gives you the advantage that will win this local war, they make proper use of their knowledge of the country (don't take tribesmen to places they do not know), and the gazelle-hunters, who form a proportion of the better men, are great shots at visible targets. A sheikh from one tribe cannot give orders to men from another; a Sherif is necessary to command a mixed tribal force. If there is plunder in prospect, and the odds are at all equal, you will win. Do not waste Bedu attacking trenches (they will not stand casualties) or in trying to defend a position, for they cannot sit still without slacking. The more unorthodox and Arab your proceedings, the more likely you are to have the Turks cold, for they lack initiative and expect you to. Don't play for safety.

23. The open reason that Bedu give you for action or inaction may be true, but always there will be better reasons left for you to divine. You must find these inner reasons (they will be denied, but are none the less in operation) before shaping your arguments for one course or other. Allusion is more effective than logical exposition: they dislike concise expression. Their minds work just as ours do, but on different premises. There is nothing unreasonable, incomprehensible, or inscrutable in the Arab. Experience of them, and knowledge of their prejudices will enable you to foresee their attitude and possible course of action in nearly every case.

24. Do not mix Bedu and Syrians, or trained men and tribesmen. You will get work out of neither, for they hate each other. I have never seen a successful combined operation, but many failures. In particular, ex-officers of the Turkish army, however Arab in feelings and blood and language, are hopeless with Bedu. They are narrow minded in tactics, unable to adjust themselves to irregular warfare, clumsy in Arab etiquette, swollen-headed to the extent of being incapable of politeness to a tribesman for more than a few minutes, impatient, and, usually, helpless without their troops on the road and in action. Your orders (if you were unwise enough to give any) would be more readily obeyed by Beduins than those of any Mohammedan Syrian officer. Arab townsmen and Arab tribesmen regard each other mutually as poor relations, and poor relations are much more objectionable than poor strangers.

25. In spite of ordinary Arab example, avoid too free talk about women. It is as difficult a subject as religion, and their standards are so unlike our own that a remark, harmless in English, may appear as unrestrained to them, as some of their statements would look to us, if translated literally.

26. Be as careful of your servants as of yourself. If you want a sophisticated one you will probably have to take an Egyptian, or a Sudani, and unless you are very lucky he will undo on trek much of the good you so laboriously effect. Arabs will cook rice and make coffee for you, and leave you if required to do unmanly work like cleaning boots or washing. They are only really possible if you are in Arab kit. A slave brought up in the Hejaz is the best servant, but there are rules against British subjects owning them, so they have to be lent to you. In any case, take with you an Ageyli or two when you go up country. They are the most efficient couriers in Arabia, and understand camels.

27. The beginning and ending of the secret of handling Arabs is unremitting study of them. Keep always on your guard; never say an unnecessary thing: watch yourself and your companions all the time: hear all that passes, search out what is going on beneath the surface, read their characters, discover their tastes and their weaknesses and keep everything you find out to yourself. Bury yourself in Arab circles, have no interests and no ideas except the work in hand, so that your brain is saturated with one thing only, and you realize your part deeply enough to avoid the little slips that would counteract the painful work of weeks. Your success will be proportioned to the amount of mental effort you devote to it.

Appendix B

"Ten Commandments for KMAG Advisors,"* 1953

As Advisor to a ROK Army Unit, I will:

(1) Take the initiative in making observations and rendering advice. Without waiting to be asked, I will give advice for such corrective actions as I would take if I were the unit commander.

(2) Advise my counterpart forcefully, yet not command his unit.

(3) Follow up to ensure that advice has been acted upon. If it has not, take it up with next higher KMAG-ROK Army Echelon for decision and action. (In ROK Divisions with US Corps, take up with the US Corps Commanders.)

(4) By sound advice and follow-up:

(a) Develop fully the combat power of all units of the command.

(b) Coordinate and control elements of the command so as to gain the greatest effectiveness in destroying the enemy.

(c) Restore promptly any part of the command which may have been lost or destroyed.

(d) Recognize battlefield conditions which might damage the potential of the command.

(e) Ensure efficient use of supplies and equipment furnished the command.

(5) Keep abreast of the tactical situation by frequent personal contact with all units of the command, using the presence of myself and my counterpart to motivate the troops and give them confidence. A minimum of my time will be spent in the unit command post. (This applies particularly to Senior Advisors and G2, G3 Advisors.)

(6) Give special attention to the training of Reserve elements, with emphasis on realism and correction of deficiencies developed during combat.

(7) Report all tactical information promptly to the next higher KMAG level regardless of reports initiated through ROK Army channels.

---

*Alfred H. Hausrath, *The KMAG Advisor: Role and Problems of the Military Advisor in Developing an Indigenous Army for Combat Operations in Korea* (Chevy Chase, MD: The John Hopkins University Operations Research Office), February 1957, 15–16.

(8)   Report deficiencies promptly to the next higher KMAG level; follow up on necessary corrective action. (Corps Senior Advisors will keep Chief, KMAG, personally informed of existing deficiencies and necessary corrective action within their purview in order that failure may be prevented rather than corrected.)

(9)   Devote particular attention to the welfare of the individual and to the maintenance of high morale and professional standards in my KMAG Detachment.

(10) Be responsible for good order, discipline, housekeeping and efficiency, not only in my own Detachment, but in all KMAG Detachments advising ROK elements subordinate to the command I advise.

I realize that I stand or fall with my counterpart. I share in the credit for his successes and in blame for his failures.

Appendix C

"Role of the Individual,"* 1962

1. General

Specific duties of Advisors by job title are outlined in Section B. . . . This section covers the role of the Advisor from a practical, day-by-day viewpoint. Some points expressed here may be more pertinent to one Advisor than to another who is assigned an entirely different job. Most of these are so indicated. Some points covered here may be repetitious or may cover "old ground" for more experienced personnel. The main purpose of this section is to offer advice to those who need it and to renew the thinking of personnel more experienced in this field.

2. The Role

a. The Advisor performs as an individual and as a member of a team.

b. As an individual, the Advisor takes upon himself the responsibility to get out and see what can be done to accomplish his mission, i.e., He generates most of the work toward that end through his own efforts. He tries in every way to advise and assist the Vietnamese so that they accomplish their work in a proficient manner. He is concerned about every problem that the supported unit or activity has, whether it be in the field of administrative and tactical operations of a Military unit or whether it be morale, living conditions in dependent quarters, pay and allowances, postal service, or general sanitation. Even if the responsible Vietnamese does not recognize the problem, still the Advisor must attempt to improve conditions which hinder or prevent the unit or activity from attaining combat readiness.

c. As a member of the MAAG Team, the Advisor diligently and willingly carries out the tasks assigned by superiors, abides by policy and guidance of superiors, and keeps superiors informed about his work. It is his duty to inform his superior when things are not going too well, or when assistance from higher echelon is required. He coordinates with other Advisors, units, or agencies when they are involved, thus completing the teamwork.

3. Rules of the Game

a. Establish good relationship with the unit or activity supported and the counterpart that you advise.

---

*Extract from *Book of Instructions for US Military Advisors to 42 Tactical Zone III Corps South Vietnam*, US Army Section, Military Advisory and Assistance Group, Vietnam, 1962, IV–C–1 through IV–C–4.

(1) Develop a genuine interest in the welfare, customs, ethics, and beliefs of the Military and Civilian community.

(2) Demonstrate your desire to pitch in and help to get things done—don't mind getting your hands dirty.

(3) Volunteer to assist in every way possible on "Do-it-Yourself" projects, such as:

(a) Design and construction of Ranges for small areas, automatic weapons, mortars and grenades (particularly anti-guerrilla trainfire-type ranges for remote CG/SDC units and training sites).

(b) Design and construction of training areas for teaching principles of ambush, counter-ambush, raids, patrols, compass courses, PBX circles, etc.

(c) Prepare and conduct training. In this regard, it must be emphasized that many Advisors, particularly those who go to remote CG/SDC sites, will be without benefit of interpreter. It is even more important to develop and utilize training aids so that the Advisor can communicate ideas and teaching by demonstration. Such devices—sketches, models, match-stick layouts, cardboard mock-ups, sand table layouts, demonstrations with personnel and/or equipment—can accomplish required training.

(d) Prepare prototypes to sell ideas for improving living conditions, e.g., improvised bath shower, pit latrine, waste disposal sump, water purification device, etc.

(4) Demonstrate and assist in supervision of maintenance of weapons and equipment.

(5) Participate in athletics. Teach new games. Learn old games. Help build facilities for sports and assist in obtaining athletic equipment.

(6) Assist in design of Security system and in improvement and execution of Security plan. Report to next Senior Advisor when required Security is beyond local capability.

(7) Avoid offending Vietnamese by showing dislike for their food, their customs, their way of life in general.

b. Dig into the status of the unit or activity supported so that you find out those things you need to know in order to give constructive advice and assistance.

(1) Determine status of personnel, facilities, equipment, weapons, ammunition and ammunition storage, communications, state of training, administration, and morale.

(2) Determine what has been done or needs to be done to remedy shortages or to secure necessary resources—follow through Vietnamese channels and keep next higher Advisor informed on assistance required by MAAG personnel.

(3) Develop an intimate knowledge of the organization, chain of command, communication system, source and system of supply, and intelligence net.

(4) Determine the needs of the Military unit or activity supported and its dependent community from the standpoint of medical, adequacy of housing, food, and clothing, etc. Work toward self improvements. Submit recommendations for assistance in improvement of conditions to next senior when such assistance is beyond local capabilities.

c. Find out what you can do to aid the local populace under the Civil Actions Program, through things you as Advisors can do or through actions the unit or agency you are advising can put into effect.

* * * * * * *

d. Take care of yourself

(1) Take care of your body; you only have one.

(2) Make your living quarters as sanitary as you can.

(3) Be alert mentally and physically, always think of security measures and fit them into the accomplishment of your mission.

e. Adhere to chain of command.

(1) Become thoroughly familiar with RVNAF channels, teach and sell chain of command to the unit or agency you are advising and assist them in effecting coordination and obtaining support through proper channels.

(2) Strictly adhere to advisory channels. Do not make promises or otherwise go out on a limb offering support, such as helicopters, on your own initiative. Senior Advisors will support you insofar as their resources allow, but you must coordinate with them or their authorized representatives before you commit such support.

f. These are observations from the present Senior Advisor, III Corps, and his principal assistants, the Senior Advisors of the three Inf Div / TZ's.

* * * * * * *

Appendix D

Advisor "Do's and Don'ts,"* 1962

Do's

Maintain your sense of humor, you will discover that the Vietnamese also possess a sense of humor.

Approach the subject under discussion from a different direction and with different words until you know that your ideas are understood.

After "planting" an idea, let the Vietnamese take the credit if it is accepted and put into practice—your satisfaction is in the net overall result obtained.

Keep abreast of what is going on in the unit, keep in close contact with commanders and staff officers to obtain information, and constantly follow-up on leads obtained.

Transact business directly with your counterpart. Do not permit the commander or his staff to subordinate you or your position.

Keep your personal appearance and conduct above reproach. Remember that the Vietnamese are careful to follow correct protocol at ceremonies and social events.

Keep a running account of major events—this is useful when it is necessary to render reports, establish the history of a subject, or follow-up. A good filing system is a must, a suspense system is also essential.

Accept invitations to Vietnamese dinners, cocktail parties, ceremonies, etc. You will find that most of them are considerate and understanding as to menus and drinks served, by exercise of reasonable precaution your health will not suffer.

When using interpreters, speak in phrases and short sentences; do not expect the interpreter to remember long speeches. Have it clearly understood with the interpreter that he will ask you and/or the Vietnamese to repeat what has been said rather than to translate incorrectly. On written translations from English to Vietnamese, always have the interpreter read back from his Vietnamese copy to ensure correct translations, remember, the Vietnamese vocabulary is limited.

Study your counterpart to determine his personality and background;

---

*Extract from *Book of Instructions for US Military Advisors to 42 Tactical Zone III Corps South Vietnam*, US Army Section, Military Advisory and Assistance Group, Vietnam, 1962.

exert every effort to establish and maintain friendly relationships; learn something about the personal life of the Vietnamese with whom you work, and demonstrate your interest—it pays dividends.

Always exercise patience in all your dealings with your Vietnamese counterpart. Never expect the job to be done at the snap of a finger.

If you find it necessary to make a suggestion or recommendation which might imply criticism of existing Vietnamese policy or procedures, do so in private, never in [the] presence of superiors or subordinates of the Vietnamese Commander.

Appreciate the work-load of the Vietnamese Commander. He will be unable to spend the entire day with you although he will probably never call this to your attention. Make yourself available at all times but let him have sufficient time to run his unit and do his paper work.

Respect the Asiatic custom and desire of "saving face."

Local conditions involving the national economy, customs, and educational development often dictate procedures which are considered inefficient and uneconomical in our Army. Avoid an arbitrary attitude toward these procedures. Try to understand them before recommending changes.

Maintain the same moral and ethical standards in Vietnam as you would at your home station in the United States. Moral degeneracy and weakness are indications of national decadence to the Vietnamese.

Try to anticipate the Vietnamese problems that your counterpart cannot foresee because of inexperience, and appraise him of the situation in time so that he can make proper and timely decisions.

At every opportunity stress the Chain of Command and its use by commanders at all echelons.

Stress at every opportunity maintenance of equipment, supply consciousness and the filling of school quotas with qualified personnel.

When advice is rendered, be sure you are on firm ground and be certain that it is within the capability of your unit to carry it out.

Be truthful in everything you say and do, as the Vietnamese appreciate and admire one who speaks the truth.

Encourage your VN associates to widen their horizons by explaining US customs, by lending US magazines, and by discussing world affairs.

Use your English classes to put your ideas across. For instance, Field Manual 22-10, *Leadership*, is an excellent textbook for intermediate reading.

Make a special effort to keep physically fit.

Always praise at least some part of what the Vietnamese do or plan to do. Then if you have criticism, couch your suggestions in tactful terms as a modification to their plans.

Set a good example in dress, posture, conduct and professional competence for the Vietnamese officer.

Present your suggestions carefully, in detail, with adequate reasons. The statement that the United States Army does a certain thing a certain way is not generally sufficient for the Vietnamese to be convinced that way is the best.

Continually stress the mutual advantages of good military-civilian relations to avoid the pitfalls of military arrogance which easily irritates the civilian populace.

Constantly encourage the strengthening of unit esprit de corps. This may well sustain the unit in the face of other difficulties.

Be able to explain or discuss basic US policy. Continually formulate in your mind how you will answer inevitable questions on current topics of the day such as racial integration, etc. However, be careful to avoid being drawn into a heated argument on the subject.

Encourage initiative and inventiveness by all commanders and officers. This trait is especially valuable in an Army of a "have not" nation that can never expect to receive all the outside material support it wants and needs.

Participate actively in the military, social and athletic functions of your unit.

Avoid underestimating the ability and capability of the Vietnamese officer. He may not have the Benning-Leavenworth [US Army] touch but he knows his own country and terrain and has been fighting on it for centuries. He and his men can be formidable opponents on their own home grounds.

Treat the Vietnamese with whom you work as you would a fellow American—equal in every aspect.

Always remember you are an advisor and have no command jurisdiction.

Shake hands with all Vietnamese officers in a room when entering and leaving.

Compliment officers concerned if and when a good piece of work is accomplished.

Exchange amenities with Vietnamese officers prior to discussing official matters.

Request copies of directives issued by the commander subsequent to your submission of a recommendation to determine if your ideas are being bought and promoted. A responsive commander will not hesitate to publish a good idea over his signature block.

Request copies of all training schedules, summaries of activities, operational reports, etc., to maintain close contact with what your unit is doing. This is also an excellent opportunity to evaluate the effectiveness of the various staff officers and detect areas of strength and weakness.

Observe your contemporaries carefully, particularly those in key slots or with whom you have constant contact, for any indications of hostility or resentment. Such individual feelings exist and your problems are compounded.

As time progresses you may think or feel that you are doing all the "bending over backwards." If you observe carefully you will find that this is not so and that the VN are meeting you halfway.

Stress teamwork and coordination.

Emphasize the importance of doing things on time by being punctual yourself. Many VN have a very casual attitude toward time.

Take every opportunity to visit other parts of VN. A knowledge of the terrain will help you understand VN military problems and will be invaluable in case of war.

Show an interest in VN customs, language, history and people. Your ideas will be more readily accepted if you show an understanding of theirs.

Keep US officers at higher levels advised of conditions of which you are aware.

Keep in mind the seriousness and urgency of your mission.

Develop the commander's efforts toward organized troop discipline.

Develop a recognition of the importance of sanitation and police.

Teach by example wherever and whenever possible.

Maintain close contact with the Commander to whom you are advisor. Tactful aggressiveness on the part of the advisor is essential.

Use highly qualified interpreters on important matters.

Persuade VN personnel to pass information automatically—up, down, and laterally.

Poke around in corners and buildings and you will find usable equipment and training aids that have not been used or even made available.

Spend maximum time in your units so that the troops get to know you and trust you.

Encourage staff officers to get out and see other regiments, other battalions train. They can't have all the good ideas. This should encourage a better parent unit esprit as well as better training.

Develop a sense of responsibility toward the unit being advised to the degree that you can feel a personal gratification for a job well done.

Try to instill, through a progressive program, US methods and practice approved by ARVN.

Consider the age and experience of commanders and staff officers at each echelon.

Think—be imaginative. The lack or absence of initiative and imagination is the only deterrent to a successful tone as an advisor.

Don'ts

Don't forget for a single minute that you may have to go to war with your unit. Any opportunity for preparation lost now may be fatal in case of war.

Don't relax your standards even though the VN standards may be lower and you are far from US supervision. However, don't flaunt your higher standard of living.

Don't forget that the VN are basing their opinion of 160 million Americans on even your most casual words and actions.

Don't forget that a careless word or action of yours can cost the US very dearly in goodwill and cooperation which has been built up here at the cost of billions.

Don't hesitate to point out faults, especially when they pertain to neglecting the welfare of the troops or wasting US aid.

Don't assume that the US school solution is the only one for VN.

Don't try to sell a US method with the sole argument that it is US. An explanation of the advantages will be more effective.

Don't condone a VN officer's attitude that officers are a privileged class without equivalent responsibilities.

Don't stimulate VN appetite for more intricate and complicated equipment by boasting about superior equipment available to US units.

Don't underestimate the VN people. They achieved independence from both the French and the Communists against incredible odds.

Don't lose a single opportunity to learn about SE Asia, especially guerrilla fighting and security in rear areas. It will be valuable to you the rest of your military career.

Don't be discouraged. Suggestions and advice you have given may appear to have been disregarded and then be implemented.

Don't drink to excess in Vietnamese company. They are a people who use alcohol moderately.

Don't ridicule the Vietnamese in conversation with other Americans. Many Vietnamese understand much more English than they admit.

Don't refuse invitations to quasi-military functions. The presence of American Advisors adds prestige to many occasions.

Don't summon a Vietnamese by shouting, whistling or hand motions. Catching the individual's eye and a head gesture will produce more effective results.

Don't criticize an individual in the presence of other Vietnamese. Always use private constructive criticism.

Don't discuss Vietnamese politics with Vietnamese personnel.

Don't fail to recognize military courtesy. Vietnamese personnel render courtesies to officers in a variety of ways unfamiliar to Americans.

Don't take offense at what sometimes appears to be abruptness and even actual discourtesy at times, it is part of the job to overlook these attitudes while at the same time doing everything possible to create goodwill and mutual understanding.

Don't accept a "yes" answer at its face value, "yes" may mean only that the person to whom you are talking understands what you have said, but it may not indicate that he "buys" your suggestion.

Don't expect the Vietnamese commander (or staff, officer) to accept all of your suggestions; he is the commander, not you.

Don't present too many subjects at one time or prolong unnecessarily the discussion of any one subject, it is better to have another conference at a later time.

152

Don't make promises which you cannot or should not carry out.

Don't show an air of superiority—regardless of rank of officers dealing with.

Don't endeavor to give advice until you have made friends with the Vietnamese officers.

Don't be afraid to get your hands and clothes dirty when giving advice in the form of a demonstration.

Don't compare relative pay scales of the American and Vietnamese Army.

Don't give advice that conflicts with directives from higher echelons of command.

Don't do the job yourself, persuade the VN individual responsible to do it.

Don't let VN personnel substitute your chain of command for theirs.

Don't hesitate to begin a project because you won't be in Vietnam long enough to complete it. Get it started and sell your successor on completing it.

Don't give up your efforts to analyze training because it is conducted in Vietnamese; get an interpreter and find out all the details.

Appendix E

"Tips to Advisors,"* 1966

*   *   *   *   *   *   *

a. Professional Duties and Interests:

(1) Sell in-place training once units return to posts. One-thousand-inch (approximately 25 meters) firing ranges are ideal for small posts to fire weapons.

(2) Spend a maximum time in your units so that the troops get to know and trust you. Keep abreast of what is going on in the unit, and keep in close contact with the commander and staff.

(3) Encourage frequent command inspections by the commander. Many often show a reluctance to inspect, relying solely on correspondence and reports to evaluate the effectiveness of the unit.

(4) Continually stress mutual advantages of good military-civilian relations to avoid pitfalls of military arrogance, which easily irritates the civilian populace. The development of a proper soldier-civilian relationship is civic action at its best.

(5) Constantly strive to raise the standards of your unit to your standards. Guard against lowering your standards to those of the unit you advise.

(6) Keep training standards high enough so that the unit is ready for an inspection at all times. This saves the wear and tear of preparation for inspection and the disappointment that follows when it's cancelled. Do not use training time for housekeeping matters; discourage the idea that the two of you can conspire to "eyewash" instructors.

(7) MACV advisors should have sufficient knowledge of all aspects of US aid programs to counter insurgent propaganda depicting this aid as interference in the affairs of the people.

(8) Constantly observe for signs of fatigue. There is a marked difference between American and Vietnamese stamina. Pushing at peak performance will cause a long-term decrease in efficiency.

b. Techniques:

(1) An advisor must constantly bear in mind that he is an advisor

---

*Extract from *MATA Handbook for Vietnam*, US Army Special Warfare School (Fort Bragg, NC: US Army Special Warfare School, January 1966), 211–216.

and not a commander. He is not in Vietnam to fight or to lead troops.

(2) Avoid rushing your acceptance by your counterpart. Overselling yourself will arouse suspicion and delay acceptance. Time spent developing a healthy relationship will pay large dividends later on.

(3) Advising works both ways. Set an example for your counterpart by asking his advice; you will get many good ideas from him.

(4) Avoid giving your counterpart the impression that each time he sees you, you are interested in asking for status reports, etc. You will soon find him avoiding you and information increasingly difficult to get.

(5) Transact important business directly with your counterpart to assure full understanding of difficult subjects. Work from the soft sell to the request for official information.

(6) Don't present too many subjects at one time or prolong unnecessary discussion of one subject; it is better to have another conference at a later time. Don't speak rapidly or use slang. By the same token, don't speak too slowly; it will insult his intelligence.

(7) Correct the most important deficiencies first. When you arrive you will see many things you will want to correct immediately. At all costs avoid the impression that everything is all wrong. In some cases it may take a month or more to sell one idea.

(8) Avoid making recommendations that lead to decisions. Leave sufficient room for your counterpart to exercise his prerogative. One of his greatest fears is that he will appear dependent upon his advisor to his troops. Carefully choose a time and a place to offer advice.

(9) Use your subordinate advisors to lay the groundwork for new ideas at their level.

(10) For successful combat operations do your homework thoroughly. The amount of advising done during combat operations is small. The advisor does most of his advising in the preparation for combat, basing his advice upon his observations or those of his subordinates during past operations. Hold a private critique with the commander upon completion of an operation.

(11) Don't be afraid to advise against a bad decision, but do it in the same manner you would recommend a change of action to an American commander for whom you have respect and with whom you work daily.

(12) Approach the subject under discussion from different directions and with different words, until you know that your ideas are

understood. The Vietnamese seldom admit that they do not understand. Don't accept a yes answer at its face value; yes may mean that the person understands but does not mean that he buys your suggestion. It may also be used to cover a failure to understand.

(13) Always exercise patience in your dealings with your Vietnamese counterpart. Never expect the job to be done at the snap of a finger—and don't snap your fingers.

(14) Information from your counterpart cannot be accepted in blind faith. It must be checked discreetly and diplomatically, but checked!!

(15) After planting an idea, let the Vietnamese take credit for it as if it were his own idea.

(16) Advisors are transient—especially infantry battalion advisors. Try to learn what your predecessor had attempted and has or has not accomplished. Ask for his files. Debrief him if you have the chance.

(17) Begin preparing a folder about your advisory area and your duties as soon as you report on the job. By posting a worksheet-type folder during your tour, you will better understand your job and your successor will have a complete file to assist him in carrying out projects you initiate.

(18) Your supervisor at the next higher echelon will often be unable to visit. He will travel with his counterpart and not get a good chance to talk with you. Your efficiency report will probably be based largely on your reports. Consider writing at least on a weekly basis to your chief. Tell him what your area is like, what [you] are trying to do, what you have been able to accomplish, what you need his help on at his level. Send him copies of advisory recommendations. Write up ideas you have for winning the war or any part of it. Your writings may give people a better idea of what kind of job you are doing. You might come up with a key solution to a problem.

(19) Take time to brief supporting pilots. Take helicopter pilots along on command visits. Try to get helicopter and observation pilots included at operations briefings. Pilots are branch qualified officers and warrant officers; they are more effective when they know the overall situation. They are less apt to complain about how they are being used when they are fully briefed on your plans.

(20) Use proper radio procedure. Your division advisory team publishes its own SSI and SOI. Remember that much advisory FM radio traffic is air-ground communication. The Viet Cong are capable of intercept!

c. Personal Attitude and Relations:

(1) Getting accustomed to the native food and drink presents a problem in somewhat varying degrees to the advisor. You will not lose face if you eat and drink with your counterpart; conversely, you will gain face.

(2) Don't become discouraged. All of your advice won't be accepted. Some of it will be implemented at a later date.

(3) Don't forget that a careless word or action can cost the United States dearly in good will and cooperation, which have been built up with great effort and at considerable cost.

(4) Don't discuss Vietnamese policy with Vietnamese personnel. It is your obligation to support the incumbent government just as you do your own. This is US national policy.

(5) Study your counterpart to determine his personality and background, exert every effort to establish and maintain friendly relationships. Learn something about the personal life of the Vietnamese with whom you work and demonstrate this interest.

(6) Set a good example for the Vietnamese in dress, posture, and conduct as well as in professional knowledge and competence.

(7) Emphasize the importance of doing things on time by being punctual yourself. Many Vietnamese have a very casual attitude toward time.

(8) Develop a sense of responsibility toward the unit being advised to the degree that you feel a personal gratification for a job well done. Do not become so involved with the unit that you cannot readily recognize failures.

(9) Accept invitations to Vietnamese dinners, cocktail parties, and ceremonies. Shake hands with all Vietnamese in a room when entering and leaving. Exchange amenities with officials before discussing business matters.

(10) Don't summon a Vietnamese by whistling or shouting. You will note that Vietnamese summon each other by a wave of the hand, similar to our farewell wave.

(11) Don't fail to observe and recognize military courtesy.

d. Personal Qualities and Requirements:

(1) Based upon observation and experiences, US advisors returning from the Republic of Vietnam have pooled their thoughts on what it

takes to be an effective advisor. No doubt each one of us is most anxious to do our best in assisting our Vietnamese allies expel insurgency from their country as soon as possible. For this reason we feel that you will welcome the opportunity to examine what other advisors have said on the subject of advising. Give these comments consideration and, to the extent indicated by introspection, make them a part of your personal attributes before and during your tour in Vietnam. These qualities and requirements, along with a general summation of desirable advisor traits, are set forth in the following paragraphs:

(a) Persevere in implementing sound advice; exercise patience and tact; display a pleasing personality; be adaptable to environment and changing situations; be honest; maintain high moral standards; be understanding and sincere; present a sharp military appearance; evince devotion to job assignment; keep in good physical condition; acquire ability to demonstrate effectively; know your job; know thoroughly the unit you are advising as to organization, equipment, and tactics; know thoroughly your own branch and have a good working knowledge of other branches; know your counterpart's problems; and demonstrate your awareness of them to him.

(b) Advisors are restricted in their operations because they are not authorized to exercise command in accomplishing advisory functions. They must rely on their ability to sell the most indefinite commodity which is represented in the individual himself. The traits of an advisor encompasses all the traits of leadership plus the ability to adapt to his environment. This environment changes with the locality or area in which the advisor is assigned. In the Far East, he must remember that arrogance and dogmatism are all the more taboo, for the religious and philosophical background of the Asian strongly opposes this type of personality. To sell one's self, you must prove your value—an advisor must present a favorable personality in the eyes of his counterpart. This can be accomplished in due time by a gradual demonstration of your capabilities in an unassuming but firm manner. Be positive but not dogmatic in your approach to any subject; however, if you are not sure of the subject matter, it is better to say so and take timely measures to obtain the correct information. To attempt to bluff through a problem will only result in irreparable loss of prestige.

(c) A most favorable trait is persistence, tempered with patience. If a problem area is discovered, continue efforts to solve it, recommend appropriate measures to be taken, and then follow through; again, remembering that patience is of utmost importance. But, the matter must be continually brought to your counterpart's attention until he is sold on taking the measures necessary to solve the problem or correct the deficiency as the case may be. The ultimate in good advising is to advise your counterpart in such a way that he takes the desired action feeling that it was through his own initiative rather than yours.

(d) Possibly the most desirable traits that you can possess as an advisor are knowledge of the subject, ability to demonstrate your capabilities in an unassuming but convincing manner, and a clear indication of your desire to get along and work together with your counterpart and other associates; however, not to the extent of obsequious behavior nor acceptance of abusive treatment. These traits, along with leadership ability and desirable character traits accepted in our own society, will usually lead to a successful and satisfying advisory tour.

*   *   *   *   *   *   *

Appendix F

"Counterpart Relationship,"* 1967

*   *   *   *   *   *   *

(1)   The advisor does not command his counterpart's organization.

(2)   He should study his counterpart's personality and background, and exert every effort to establish and maintain friendly relationships.

(3)   The advisor should make "on-the-spot" recommendations to his counterpart, when appropriate.

(4)   The advisor may represent his counterpart in disputes with US agencies; however, this representation should be based on sound judgment and not blind support.

(5)   The advisor must not present too many subjects at once or prolong unnecessarily the discussion of any one subject. Suggestions and recommendations must be within the counterpart's capability to carry them out. Avoid harassment.

(6)   The advisor should never accept "yes" at its face value; "yes" may mean only that the person understands what has been said (it also may be used to cover the failure to understand), *not* that the counterpart "buys" the recommendation.

(7)   The advisor should present recommendations carefully, in detail, and supported adequately with an explanation of advantages inherent to the proposal. Recommendations which require immediate decisions should be avoided, except when the situation dictates. Counterparts should be allowed to exercise their prerogatives; one of their fears is that they may appear overly dependent upon advisors. The advisor should choose appropriate times and places to offer advice.

(8)   The advisor should not convey the impression that everything is all wrong. A careless word or action on the part of the advisor can impair the advisory effort. If there is criticism, it should be couched tactfully, but the advisor must not be reluctant to criticize when criticism is in order. Failure to do so may leave the counterpart with the impression that the advisor does not know or care. Appropriate, timely, and tactful criticism can engender respect. If it is necessary to make a recommendation which

---

*Extract from Department of the Army Field Manual (FM) 31–73, *Advisor Handbook for Stability Operations* (Washington, DC: Headquarters, Department of the Army, October 1967), 51–54.

might imply criticism of HC policy, advisors should do so in private.

(9) The advisor should ask the counterpart's advice; he has many good ideas. The advisor who tries to oversell himself may arouse suspicion and delay acceptance. Do not make promises which cannot or should not be fulfilled.

(10) A subject should be discussed until it is known that the counterpart understands.

(11) Frequent inspections should be encouraged. It may be necessary to convince the counterpart of the value of frequent inspections to determine actual conditions.

(12) Initiative and inventiveness should be encouraged. The counterpart may follow orders to the letter and, even if a modified course of action subsequently appears to be more appropriate, he may not deviate (or request permission to deviate) from his original instructions. The advisor should encourage his counterpart to request changes in orders when the need is obvious. Encourage him to be receptive to such requests from his subordinates.

(13) A project should not be rejected because the advisor will not be in-country long enough to complete it. Major events and projects should be documented and transferred to successors. Briefings, end-of-tour reports, and other instructions will assist in providing a smooth transition and continuity of effort.

(14) Maintain a filing and suspense system. Secure classified documents.

(15) Definitive goals and objectives should be developed as part of the overall advisory program. Systematic evaluation ensures continuity of advisory effort.

(16) The advisor should keep abreast of activities and in close contact with civilian political leaders, military commanders, and staff officers.

(17) The advisor should participate actively in military, social, and athletic functions. If unable to accept a social invitation, regrets should be expressed in accordance with the local customs. Invite counterparts to appropriate social functions.

(18) A sense of identity with the counterpart's unit or area should be developed. Spend maximum time at the scene of activity. Attempt to learn the language and volunteer to teach English.

(19) Subordinate advisors should lay the groundwork at their levels for new ideas.

(20) The consequences of mistreating suspects or prisoners should be stressed. Captured insurgents and other persons taken into custody should be treated humanely. The minimum requirements for humane treatment are specified in Article 3 of the Geneva Convention and include: Care for sick and wounded; prohibiting violence such as murder, mutilation, cruel treatment, and torture; taking of hostages; outrages upon personal dignity such as humiliation and degrading treatment; and the passing of sentences and carrying out of executions without previous judgment pronounced by a regularly constituted court. Insurgent subversive elements are subject to laws concerning subversion and lawlessness. Advisors must not become involved in atrocities. They should explain to their counterparts that they must report any atrocities of which they have knowledge. Captured insurgents should be interrogated immediately at the lowest level for tactical information. The loss of a prisoner, whatever the justification, is a loss of a valuable intelligence source (FM 30-15, FM 30-17, and FM 30-31).

*   *   *   *   *   *   *

Appendix G

"Advising the RF/PF,"* 1971

*   *   *   *   *   *   *

a. Always remember that you are an advisor and have no command authority over your Vietnamese counterpart.

b. Keep abreast of what is going on, keep in close contact with your counterpart to obtain information and follow up on leads obtained.

c. Remember that information obtained from your counterpart cannot always be accepted with blind faith. It must be discreetly and diplomatically checked.

d. Don't hesitate to make on the spot corrections; however, do it tactfully. Don't walk away from something that is wrong; indicate the need for improvement while noting diplomatically those things that have been improved.

e. Advising can work both ways. Set an example for your counterpart by asking his advice and you will get many good ideas from him.

f. Always praise at least some part of Vietnamese plans. Then if you have criticism, phrase your suggestions in tactful terms as a modification of their plans.

g. Present your suggestions carefully, in detail, with adequate reasons. An explanation of the advantages will usually be effective.

h. Do not present too many subjects at one time or prolong unnecessarily the discussion of any one subject.

i. Understand your counterpart's personality, know his background, and exert every effort to establish and maintain friendly relations.

j. Set a good example for the Vietnamese in dress, posture, and conduct, as well as in professional knowledge and competence.

k. Emphasize the importance of doing things on time by being punctual yourself.

l. Emphasize the importance of doing things correctly the first time, to avoid setting a poor example or introducing poor habits.

m. Be familiar with your counterpart's experience and military

---

*Extract from *rf / pf Advisors Handbook January 1971*, Headquarters, US Army Military Assistance Command, Vietnam (RVN: Headquarters, US Army Military Assistance Command, Vietnam, January 1971), 7–9.

training. Your ideas will be more readily accepted if you show an understanding of his point of view.

n. Keep in mind the seriousness and urgency of your mission. Many things can be accomplished if you maintain a high degree of motivation.

o. Don't be discouraged. Suggestions and advice which you have given may appear to have been disregarded but may later be implemented.

p. Don't fail to observe and recognize military courtesy. If you are not of higher grade than your counterpart, treat him exactly as if he were your US senior.

q. Be aware of all problems but don't become engrossed in your counterpart's minor every-day problems. Place your advisory emphasis on the overall effort.

r. Remember, you are working with a culture in which methods and outlook are not the same as the one to which you are accustomed. Therefore, exercise patience and understanding.

    *   *   *   *   *   *   *

Appendix H

"Points for Consideration,"* 2001

This section provides a number of specific points that should be considered when dealing with foreign counterparts.

Successful advisory efforts rest largely upon human relations. Everyone knows that it is not always easy to convince people who most need assistance that they will receive any real benefits from it; therefore it is of first importance that SF soldiers help the rank and file people working with them see how they can benefit from their joint efforts.

Likewise, it is sometimes difficult for advisors to establish rapport and work effectively with foreign nationals. The following points are based on first-hand experiences in recent deployments but they match the experiences of technical cooperation programs abroad, and, before that, in rural improvement programs in the Southern United States. They are concrete suggestions dealing with the difficult matter of how the advisor can best use his skills.

Be Sure Your Presence is Understood

The advisor needs to enter the area where he will work under the right sponsorship. This will include local military authorities, but also the mayor, village head, or some other recognized local civilian leader. Prior to this, his arrival may need preparation through the district, sector and/or other appropriate local offices. No amount of clearances from the distant national or state/province/sector government can compensate for local explanations of why American Special Forces personnel are in the area. This is especially true of small and/or isolated communities where it is unusual for a stranger to appear for even an hour without being acknowledged and accepted by local leaders.

Without explanations from locally respected persons (opinion leaders), the local population will arrive at its own explanations, often to the detriment of the SF efforts. One Central American village became convinced, for example, that the Americans were there to steal children.

---

*Extract from *Special Forces Advisor's Reference Book*, US Army Special Warfare School, Research Planning, Inc., October 2001, 100–105. Reprinted with permission of L3Communications (www.titan.com). This is not a US Army manual; it was developed under contract for Commander, US Army Special Forces Command.

Find a Basis for Common Interest with the Local People

If the advisor shows appreciation of host country nationals as individuals, common ground can usually be found despite culture gaps and language barriers. He can listen when they talk, and look with interest at what they show him. Initial conversations will usually center on universal matters such as food, shelter, clothing, health and education. In time discussion can be naturally brought around to the matter the advisor wants them to consider. He will be better received if he knows something about earlier host nation military history, contributions in such matters as agriculture, folk art, religion, architecture, and so on. Naturally, he will be more effective and appreciated if he can speak the local language.

Try to Understand Why They Do Things the Way They Do

Some local practices may seem strange and non-sensible at first but they generally have good reasons behind them that the advisor can discover with good observation. A creative imagination helps. Gleaners in the Near East, for example, operate within a folk framework that gives support to the elderly, somewhat like the American Social Security program. Food habits, family traditions, folk cures, and festive celebrations nearly always have a great deal of human experience behind them. The advisor will need to be alert to the fact that many local military units or villages contain rival sub-groups and factions; he will need to reckon with these. Factionalism (in its most intense expression, feuds), in small isolated groups generally, seems to serve to lessen the monotony and boredom of everyday life.

Start Where the People Are and with What They Want

The lives of traditional peoples anywhere in the world are usually simple and realistic. It is important to find out what the local people really want most and work with them to get it. They may want a public school, or a road, when the SF team thinks the village most needs a well or a clinic. The need the local people feel may often be the best starting point, regardless of its comparative merits. Then people are more likely to be appreciative and cooperative, to begin to raise their sights and become interested in working for other improvements. To service the initial desires of the people, the advisor may need to call in other personnel with skills needed for the particular project. This can involve some delay, but this way he gets eventual full cooperation in other projects. Sometimes the desire to show immediate results causes the advisor to press for a project despite the desires of the local population. In this case he will at best get only a half-hearted response, and may put American assistance in a bad light locally.

Work within the Local Cultural Framework

The SF advisor needs to understand such basic cultural matters as the ethnic background(s) of the people, family relationships, leadership patterns, value systems, and the technological level of the people as related to ways of making a living. He also needs some knowledge of local services such as health, education, and communications (including transportation). Many things will depend upon the advisor's understanding of cultural issues—for instance, the extent that locally available physical resources can be used.

Help People Believe They Can Improve Their Situation

The vast majority of the traditional peoples of South America, Asia and Africa have long lived in a more or less static situation. This can even affect the local military. Through experience, they have come to be more fearful of losing status through change than hopeful of bettering their condition through change. Therefore, a suggested change is often viewed with fear. Concrete projects that yield easily observed benefits are helpful in convincing such people that they can improve their situation and make them more willing to cooperate in other projects.

Be Content with Small Beginnings

First changes nearly always come slowly in areas where there have been few in recent times. It is good to remember that, historically speaking, scientific development in the West occurred only recently. The advisor should keep in mind that knowledge, whether technical or otherwise, is cumulative, and that once a small beginning has been made, greater activity and changes will likely follow. But remember, it is easier to achieve momentum than it is to maintain it. The important thing is to make a start, within as promising a framework as possible, and with the support needed to sustain the momentum achieved.

In some areas, people may have had the experience of various assistance programs that upset their traditional way of life, but provided no lasting benefit. This can also make them suspicious of outside "assistance."

Utilize Local Organizations and Recognize Their Leaders

People everywhere respond best when their local organizations are recognized as important and useful. A program is unlikely to succeed unless it is carried forward within the local organizational framework. The recognized local leaders, military and civilian, must be consulted and encouraged to make such contributions as they can. A well-conceived technical activity will reflect credit on the local leaders associated with it. But also, attention must often be given to the quiet, behind-the-scenes leaders, no less than to ranking military officials and the family heads of local groups. The surest way for an activity to be continued after the SF team leaves is for it to have been launched and carried forward within the local organizational and leadership framework.

Help the Government Get Organized to Serve the People

For the advisor to be most effective he must understand not only the local military but also the local government set-up, and how his activity fits into the overall scheme of things. There should already be a set of agreements between the military, various local national agencies and government Ministries, usually through some sort of inter-ministerial council that provides for coordinated effort in servicing the varied needs of the local people. The SF leadership should work with appropriate agencies to assist in getting such agreements made. If such agreements already exist, SF personnel should be careful to recognize and strengthen them. The work of the advisor in one field is most meaningful when properly coordinated with the contributions of individuals and agencies in other fields.

Train and Use Semi-Professional, Multi-Purpose Local Workers

Selected young people in the villages can be trained and used as sub-professional, multi-purpose village workers to enable the advisor to make the best use of his time. Otherwise, the SF soldier's influence is restricted to where he is standing and the immediate vicinity. Furthermore, the advisors, and especially their leadership, spend much of their time establishing and maintaining rapport. The gap between the local population and the SF advisor is usually a formidable one because of the great educational and cultural differences between them. Often the team works with villagers who are poor, illiterate, and often have experienced few or no outside contacts. Volunteer or paid local workers (who serve as liaison between the villagers and the advisors) have proven of great help in getting the benefits of subject-matter technical activities. There are young people in the cities and villages of South America, Asia and Africa who are eager to be trained and used in local developmental activities. The training of a local worker is two-fold: to teach him (or her) the many simple things he or she can do to help the villagers help themselves; and to help him or her understand what they cannot do, and how to call in advisors as needed.

Expect Slow Progress

As the local people, on the basis of their own successes from their joint efforts, begin to have new hope they naturally want a larger hand in matters. The advisor may sometimes feel they want to assume more responsibility than they are able to carry. These evidences of growing pains should be appreciated, for they are a necessary part of becoming able to assume responsibility. They indicate that the local population is beginning to believe they can do more and more things for themselves. The advisor needs to adjust himself to these growing desires of the people to help themselves.

Transfer Controls Constructively

The matter of institution building is a challenge to the SF team. They need to help the local people see how they can build the new (that they want) upon the foundations of the old (that they already have). From the beginning of a project the team needs to have envisioned, at least roughly, and to have discussed with local leaders, the various types of training of local personnel needed, the means by which needed financial support can be had, and the several progressive transfers of responsibilities that are to be made before the full operation of the activity can be relinquished. How can the team know the timing of withdrawal best related to local conditions such as personality characteristics, value systems, and so on? If operating responsibility is transferred too early there will likely be some breakage, as it were, usually of material things; it is important to note that the cost of such breakage can usually be charged more or less to training. If, on the other hand, the team keeps control too long, the local people who have wanted to take over may become disillusioned with them or even hate them for not relinquishing control to them when they thought it should have been turned over to them. This delayed handing over of responsibility moves the problem out of the material level and into the psychological, which is the more difficult to cope with. It must be clear to all, that the team has the challenge of working out with local leaders the timing of the phasing out of each activity so that changes can be institutionalized.

Don't Demand Thanks from the People Helped

People who benefit from assistance are seldom in a position to be grateful. Rather, they are usually aware that they are making headway belatedly, and they may tend to be on the defensive. In accepting assistance they in a sense admit their own insufficiency. A person's or community's or a nation's self-esteem is a precious thing. The team therefore should not expect thanks, but instead approach the people in a spirit of fraternity and humility, taking satisfaction in such progress as they may make and being quick to see that the credit rests with them. The team should do its job the best it can and accept work well done as is its own reward. Insofar as other monuments may be needed, mankind will, even if a bit belated, erect them in the right places and to the right people.

The full application of the above points rests upon yet another dimension, namely the need for the SF advisor to achieve a working equality between himself and the people with whom he is working. One reason this dimension is so difficult to achieve is that the advisor tends to assume he has achieved it already, when in fact he often has not adequately identified its components, many of them quite elusive. This final point warrants some detail. The relationship between the advisor and his counterpart or local people may seem like that of teacher and student. In some cultures it may be regarded as that of master and servant. All of these relationships imply a basic inequality in individual worth. And any such implication, or inference, negates the rapport needed to accomplish the very end the advisor seeks, namely the development of that greatest resource of all, the human resource.

The degree of identification between the members of the SF team and the people is a most important component in the achievement of a working equality between them. In actuality the SF advisor will be able to accept as equals only those in whom he can see himself, though under a differing set of life circumstances. He can treat the Bedouin or fellahin as an equal only when he understands that if he'd been in the same circumstances all his life he'd be making a living in about the same way, speaking his language, following his courtship and marriage customs, and responding to about the same set of fears and hopes. Such identification is not a superficial thing; it is learned through extended exposure and deep insights. The reverse, too, is important—the advisor helps the local people realize that they would be about like he is if they'd been in his situation all their lives; such identification becomes a dynamic change that leads to improvements in local living conditions. The difference between the advisor and advisee is a product of circumstances, not a question of individual worth. A really important thing happens when, through identification, the advisor "understands" the people with whom he is working; and when they, looking at him, begin to believe that they can help change their own situation.

Fortunately, the joint efforts of the advisor and the people in meeting a specific local felt need provide a basis for effectively working together despite differences in religion and value systems, despite differences in economic status and social position. This joint effort provides a framework in which the advisor can make maximum use of his supportive background (a highly developed country, with a heritage of a successful revolution, a high value on class mobility, and an interest in helping other people help themselves). Conversely it reduces the inherent handicap in the marked national differences in wealth, health, education and technology. In short, the joint effort between advisor and the local people to effect a desired local improvement—whether of a simple material type such as a pump or clinic, or a shift toward greater self-direction for the people in their own affairs—constitutes a working relationship that helps to overcome the superficial differences among men and so affirms equality, and brotherhood.